# Engineers

This innovative new book presents the vast historical sweep of engineering innovation and technological change to describe and illustrate engineering design and what conditions, events, cultural climates and personalities have brought it to its present state.

Matthew Wells covers topics based on an examination of paradigm shifts, the contribution of individuals, important structures and influential disasters to show approaches to the modern concept of structure. By demonstrating the historical context of engineering, Wells has created a guide to design like no other, inspirational for both students and practitioners working in the fields of architecture and engineering.

**Matthew Wells**, a chartered architect and engineer, has taught at various schools of architecture including Nottingham School of Architecture as a unit tutor. He is an external examiner at Liverpool University School of Architecture and has been a technical tutor at the Bartlett School of Architecture for 10 years. He is best known as the founder and director of Techniker Ltd, Consulting Structural Engineers and has worked with leading architects such as Lifschutz Davidson and Eva Jiricna on footbridges, glass staircases and many contemporary leading edge projects.

# Engineers

## A history of engineering and structural design

**Matthew Wells**

Routledge
Taylor & Francis Group

LONDON AND NEW YORK

First published 2010
by Routledge
2 Park Square, Milton Park, Abingdon, Oxfordshire OX14 4RN

Simultaneously published in the USA and Canada
by Routledge
711 Third Avenue, New York, NY 10017

*Routledge is an imprint of the Taylor & Francis Group, an informa business*

© 2010 Matthew Wells

Typeset in Univers by
Florence Production Ltd, Stoodleigh, Devon

*British Library Cataloguing in Publication Data*
A catalogue record of this book is available from the British Library

*Library of Congress Cataloging in Publication Data*
A catalog record has been requested for this book

ISBN10: 0–415–32525–0 (hbk)
ISBN10: 0–415–32526–9 (pbk)
ISBN10: 0–203–35818–X (ebk)

ISBN13: 978–0–415–32525–7 (hbk)
ISBN13: 978–0–415–32526–4 (pbk)
ISBN13: 978–0–203–35818–4 (ebk)

# Contents

# Illustration credits

Archivio Pier Luigi Nervi, PARC/MAXXI Museo Nazionale delle Arti del XXI secolo, Collezione MAXXI Architettura ©, 11.4

The Art Institute of Chicago ©, 12.0

BAE Systems via Brooklands Museum ©, 10.0

Dr Marco Baldari, Grosseto, Italy ©, 4.6

Bibliotheque Nationale de France ©, 4.3

Robert Cortright/Bridge Ink ©, 5.2

Robert Cortright/Bridge Ink ©, 5.4

Collections/John D. Beldom ©, 7.2

Collections/Gena Davis ©, 11.9

Constance Tipper: personal papers, Newnham College Archives C, 10.5

Copyrights have expired ©, 9.3

Yann Arthus-Bertrand/CORBIS ©, 1.0

Alinari Archives/CORBIS ©, 1.5

Historical Picture Archive/CORBIS ©, 1.8

G.E Kidder Smith/CORBIS ©, 2.2

Bettmann/CORBIS ©, 2.3

Owen Franken/CORBIS ©, 2.6

Galen Rowell/CORBIS ©, 3.0

Murat Taner/CORBIS ©, 3.2a

Christophe Boisvieux/CORBIS ©, 3.5

Reuters/CORBIS ©, 3.6

Kevin R. Morris/CORBIS ©, 4.0

Paul Almasy/CORBIS ©, 4.1

Christie's Images/CORBIS ©, 4.2

Bettmann/CORBIS ©, 4.5

Alinari Archives/CORBIS ©, 4.7

Nick Wheeler/CORBIS ©, 5.0

Bettmann/CORBIS ©, 6.4

Annebicque Bernard/CORBIS SYGMA ©, 6.5a

Martin Jones/CORBIS ©, 6.6

Hulton-Deutsch Collection/CORBIS ©, 7.5

Bettmann/CORBIS ©, 8.3

CORBIS ©, 8.4

Bettmann/CORBIS ©, 10.1

# Preface

For me architecture proved to be difficult and diffuse while engineering promised the crystalline and complete. Why then do I attempt to problematise engineering in these pages?

The craft of engineering has an easy elegance, the classical maths it uses is appealing and the expedients available to practice give the sense of a game well played.

However, simultaneous with the feeling that the general project of structural engineering is nearly complete comes another unease: that maybe the whole edifice has more contingency than inevitability about it. Maybe there are other equally valid ways of designing building structures. Is the consultant engineer's role petering out or instead on the verge of a new diversity?

I was subjected to an English architectural education in the early 1980s. The emphasis then was on a breadth of knowledge, science and art. Soft Scandinavian modernism was the dominant style; the English had chosen to appropriate it in a manner characterised by a pervasive lack of consensus. Attempts to systematise the design process by Leslie Martin, Christopher Alexander and others were being abandoned and dismantled. The study of these efforts became the subject of my final-year thesis.

Unfortunately, the secondary school I attended had not offered the mixture of subjects – physics, mathematics, history and English – that the Royal Institute of British Architects recommended as a suitable grounding. Lacking in Arts, I only sensed that a discussion of engineering design 'in-itself', particularly its inconsistencies, might have lasting usefulness.

Of those I have worked for, three have been inspirational. Alan Baxter, Tony Hunt and Tim Macfarlane. The complexity of thought and the contribution of each are beyond me to explain. The co-existence of these three and many other talented engineers in London at the turn of the millennium makes it a truly intellectually fertile environment in which to work.

I would like to thank my publisher, Katherine Morton, for her professionalism and patience, and Matthew Hollins, whose graphic expertise provided the diagrams. Della Pearlman made me a bibliography and together with Velia Albertoni and Elizabeth Pinches located images. Most of the ideas here have been gleaned from my work colleagues over the years.

My tendency to build castles in the air as well as earthbound is well known to those around me, and many of the more outlandish conclusions drawn in earlier drafts of this book have had to go. Those that remain are my responsibility alone and I hope that the mistakes herein prompt vociferous correction and discussion.

My current holiday reading is a potted history of the Phoenicians in the western Mediterranean. Beaten at sea by the Carthaginians, in 243 BC the ancient Romans took a captured quinquereme – a five-banked galley famous for blockade-running – replicated it two hundred times* and returned to appropriate the earlier civilisation's entire enterprise. The point is that it is not often the originators of something who ultimately benefit. It is a privilege to be there when the ideas form and it is a privilege to be able to exploit them subsequently. This two-part equation of design, the originator and the perpetrator, could be so much more fecund if discourses didn't have to extinguish one another. There is such a rich background upon which the modern engineer can build. It is not only Newton's shoulders that we stand upon.

Ibiza

*Despite copying every other detail, the Romans overlooked the need to season timber so their fleet fell to bits over the single campaign – engineering has many aspects.

# Introduction

This book started life as a sheaf of notes appended to an office design manual. Along with design codes, section tables and specifications, the prescriptions and preferred ways of doing things built up over years by a busy engineering office, it was intended as a working tool, to be set beside all the other paraphernalia for day-to-day use.

Modern building engineers train and practise within a hermetic, yet notionally universally applicable, system of rules. Well aware of the limitations of their intellectual tools, they exercise their dominion over nature with care. They may be less conscious of the background to their confidence, the alternatives that have passed away and the sediment of ideas, unconvincing in their inevitability, that influence their actions.

The purpose of this study – a collection of methods, biographical details and case studies set out in linear-historical narrative – is to act as an enabling device in the design process. How is it meant to work? Like a set of intellectual spanners loosening the firmly fixed. From within the confines of their modern and well-stocked consciousness (consciousness meaning only the constant implicit knowledge that we apply to the world), the practising engineer and interested reader are asked to consider the conceptual spaces in which our predecessors have been obliged to work. The paths along which engineering expertise has developed are traced, with many crossroads and dead ends glanced into. Abandoned ideas and marginalia are given special place, not only for their own potential and interest but also for the deep insights they offer into the ways that have ended up being adopted and canonised. What follows is recognition that our current outlook is grounded in many opaque sources and is not at all inevitable. The model of technological development as a self-sustaining trajectory is replaced by one in which accident and contingency decide between many possible outcomes. There is diversity in engineering, which is continually resurfacing and which in itself may be controllable.

The underlying assumptions of our received knowledge are drawn out to be held up in the light of rigour and enquiry. One view is that too much self-conscious reflection compromises the ability to act, and the perils of Friedrich Nietzsche's 'hypertrophy' – the inability to act brought on by a morbid preoccupation with trying to understand and to be only moral or aesthetic – seem particularly fearsome to many engineers. However, the avowed purpose of us self-declared engineers is to control our surroundings. We are born and enter a social system; our responses are coded and innate. Most of what we do and think and design is governed by an extensive structure of assumption, tradition, craft and acceptance. It would seem wise to be aware of some of the inconsistencies of the system, albeit that safeguards have formed against adverse consequences. But we are engineers, and we want to understand, or rather to manipulate, all the mechanisms of the universe. When we identify these invisible links, we will manipulate them. And why? Because we can.

An assumption is made here that human consciousness works within conceptual spaces, areas of coherence, teleologies that vary between individuals and groups and over time. This book surmises that if sophisticated structures can be conceived of and constructed by people from other times with different resources and realities, then our own boundaries may be extended by seeing things in different ways, and by remaining continually wary of the traps of preconception. For who is not aware of the mental ties that bind: what has gone before, the right way of doing things, the neat completeness of the technologies, the impossibility of doing anything but working over the palimpsest of earlier understandings?

Although the following is organised as a review of engineers, technical ideas and devices set out in chronological order, this is not to imply a linear process of development. Rather, an over-simplified description would be that it is one of layering, where successive scenes are formed over and influenced by past sequences, so that at any one time there are a myriad of processes operating simultaneously, some ancient and some deeply embedded. The method here is to follow the layering of thought to which each generation has been obliged to respond in some way, perhaps adding to it, or at the very least, rejecting it and then filling the resultant vacuum. This approach does not assume a process of revolutionary advance.

Thomas Kuhn's idea of paradigm shifts, where a more coherent system displaces a lesser one at a distinct change point, recognises that systemic continuity is an essential condition for the occurrence of change and its acceptance, and therefore major shifts involve dislocation. His thought proves less helpful in not accommodating the simultaneous and sometimes mutually reinforcing operation of independent patterns in single contexts and the occasional and happy existence of extraordinary forms inexplicable to the dominant paradigm. Technical and conceptual developments can occur smoothly as well as in abrupt and over-determined lurches triggered by external influences.

A reading of evolutionary theory sometimes seems to offer a closer model of technical change, with its series of almost randomly generated alternatives, tested by their environment. Communication ensures Lamarckian improvement, and perfect form eventually emerges. Human art and the continual recreation of culture short-circuits these notions. The Marxist idea of technical advancement as the inevitable response of the means of production to the pressure of capitalism is also incomplete. The way we engineer is in itself pure artefact and has to do completely with the personalities involved in its formulation and the historical facts of what happened around them. Competition, personal animosity, and power-seeking are revealed as key drives.

The rescue of Freudian psychology's reputation from sustained, concerted assault and revelations of deceit has been partially effected by demonstrations that the phenomena he discovered can be identified in artworks and artefacts. The great man's insights deployed here pertain not to the flawed theory on personality drives or the structure of the subconscious but to the ways that human processes may be shielded or shrouded in some way and are to be understood indirectly. In these pages engineering design is seen as an ever-widening set of complete systems, contributions made by individuals, alongside a group consciousness tracking across an intellectual space and responding, with subtle or abrupt shifts, to influences of various identifiable kinds.

Through the ordering of events and ideas, we are trying not only to identify underlying agencies but also, and most particularly, to reveal the junctures when the assumptions that we are concerned with were made, and their nature as revealed by their origins and gestation, and then to reveal an engineering that preceded such a concept's existence. For there has always been such engineering with other discourses – much of it precocious. In this investigation we are particularly drawn to obsolete engineering, to the systems, approaches and attitudes that have come and gone. Like ideologies defeated on the battlefield, they were extinguished by prevalent power structures but often remain just as interesting and valid as the systems subsequently achieving the status of received knowledge. Excursa, digressions, *never-realised*, *never-adopted-as-the-canon*, are just as legitimate as the ones that were. There are other worlds of engineering with an existence and consistency equal to our own. As the implications of Darwinism were thought through, corollaries of the missing link and evolutionary dead ends were conjured up; monsters and chimeras must have existed for the selection process to function. And there they were in the Burgess Shale,[1] fantastic creatures passed from time. We must find our own engineering 'Burgess Shale' to yield otherness.

Another reservation regarding the use of a chronology in this study draws on a literary concept: the notion that authors influence their precursors. Writing modifies its own past as well as affecting the future. The work of a successor always repositions the forerunner. Each contribution to engineering alters everything that has gone before. Can these shifts be considered physically consequential? On first sight Gustave Eiffel's Garabit viaduct does not physically change down there in that deep wooded river bend, but one's understanding of his thought alters profoundly after encountering his built work. But what is that reverberation, as each generation of visitors to the bridge sees something different before itself? In those encounters buildings yet to be built are affected. A critical study of engineering is a two-way process; alighting at any point, the observer changes what is up and down the line.

In the past many of the individuals referred to in this book have been situated in sequences of development to which they do not really belong; for a meaningful analysis, an individual's unique input must be continually realigned and reinterpreted in support of larger trends. Can the losses attendant on these oversimplifications be mitigated in some way? First, the output of each personality can instead be judged as complete within itself. While it is not possible to completely free the recording of a body of work from all interpretation, broader, unfocused descriptions do at least allow the reader the possibility of continually rebuilding conceptual models and resituating the subject under consideration. Second, the lack of fit between the 'real' individual and his idealised cipher must present clues. To what extent does each endeavour match itself to the background signal so that the two simultaneously create noise? An entropy argument appeals here. Can one's output overtly act to reinforce one's surroundings, also disrupting them and opening spaces in them for the unforeseeable to enter all at the same time?

As an outcome of this study we imbue engineering with a cultural significance, both as expression and as determinant. In the same way that the ancient Egyptians overlooked their urbanism, their art dwelling only on the religious and political because their cities were in themselves sign systems carrying meaning, so for us engineering is seen to be largely ignored by art because it is already a monument to itself. Only in the stirrings of modernism, in the Impressionist paintings of the train sheds and in Futurism, do we see the first serious attempt to criticise engineering itself, something more than the shock of the machine. A critical overview for engineering is long overdue: nothing to do with the efficiency or beauty of a structure or machine, but a nuanced assessment, continually rehearsed, of what it is that we think we're doing. We now encounter very great differences between contemporary practitioners, perhaps eroding under globalisation, and the co-existence of incommensurable technologies, sometimes in close proximity. It is the ways in which engineering is being carried out, not just its products, that are worth examining; otherwise, we should only look ahead. And that is the essential characteristic of all artefacts. Nature is coded only at the lowest level; the significance in embodiment is humanity or its lack.

Why is the focus here on structural engineering, and is that restriction of range necessary or desirable? As noted in the first paragraph of this Introduction, these researches were for use in particular practice, and anyway, there must always be a propensity to speak of what one knows. However, building is one of the very oldest technologies. Boat-building offers a good alternative, and the simplicity and accessibility of innovations in naval architecture appeal. Here and there in this essay a marine example is chosen to illustrate a particular point. Weapon-making, too, is old and so very human, which reflects uneasily on creativity. The engineering of buildings is ubiquitous and interfaces directly with all the other products of the human consciousness. Builders, designers and builder/designers are a fascinating grouping to study, the characters highly differentiated and seemingly of all types.

Engineers' ideas have had profound effects on built form. Construction has in turn been subjected to outside influences, forcing engineering itself to respond in turn. Of all the forms of engineering, the structuring of a building is the most loosely constrained by nature, with even the tallest tower or widest bridge far away from those technical limitations imposed on aircraft or electronics designers. This leeway admits other influences and can be used to explore other effects. Buildings have always reflected many preoccupations and serve as an accurate record of past enquiries if they can only be diagnosed correctly and placed in proper context.

Eleven of the chapters here run from prehistory to the most recent time when engineering was prosecuted with complete confidence, with our immediate past reserved for the final section. We have sensed among contemporary engineers consensus dissolving into a vacuum. Within the next decade the engineering of buildings could become completely automated, so what next for engineers? Does the past offer any indicators for our future?

Chapter 1 begins with the first orderings and humanity's handling of materials. From these stem some wide-ranging generalisations on the structure of the mind and the relationship of consciousness to the reality upon which we must operate. Karl Popper's thoughts on the structuring of knowledge in the West provide a baseline. We have always been trying to win an imperfect knowledge from the all-knowing gods, and once we believed understanding to be there at all and, next, amenable to ourselves, a course was set. Geometry was the first of Giambattista Vico's (1668–1774) artefacts to govern our subsequent output; it gives shape and consistency to our universe. The stars were seen to be reliable, immutable, perfectly predictable, and from this followed the realisation that the abstraction of number seems to contain axiomatic sets, properties applicable anywhere. Belief in discovery, nature following laws that can be unveiled and furthermore presumed to be representative of an inner logic, is taken for granted now.

Are these precepts, among the oldest we have, an inevitable part of our thinking? Are the most embedded structures the most powerful? Notions that these findings are inventions and that such simplifications act to obscure as well as illuminate remain disturbing. Their nature certainly makes them difficult to manage and they seem to cloak their subject as they operate. Realisations of individuality and therefore otherness, possibilities of influencing nature, the symmetrical development of mind and physical tools, social constructs and organisation, conceptual thought and abstraction are indivisible components of one whole.

For most of man's history engineering has been carried out in an unutterably different 'other way' from how it is achieved in the present. Studying this otherness, one sees it to be not primitive but rather *complete-in-itself* and timeless. In parallel with a different literary consciousness, there was a relationship with nature alien to our own, so is this still with us to be activated or drawn upon in different ways?

Nature's building materials appear to be very varied. The range of properties to be manipulated by early man was extraordinarily wide, seeming to us to meet each need as it occurred. The body of a Neolithic shepherd, ambushed, mortally wounded and fleeing to die alone on an Alpine glacier, was preserved and recovered five thousand years later.[2] He was found still carrying a full rig of equipment made from a dozen materials. Various woods had been carefully selected for different purposes: elastic yew for an axe, stiffer larch and hazel to fashion a backpack.

The intellectual abstractions of geometry permitted codification and increasingly complex prescriptions for structure to be created, controlled and communicated; the pure form of music or song was used to convey design information. In Chapter 2 the ancient world's explosion of information, recording and interfacing is considered. 'Meta-artefacts', devices to govern the understanding, describing and making of things, rapidly gained sophistication and the so-called early civilisations, despite their variety, seem to have exploited linguistic and literary systems in similar ways. Ungovernable economic forces appeared alongside urbanism. Notions of efficiency, and therefore attitudes to standardisation, specialisation, special display and sign systems, took shape. Attendant on the development of social structures and mental artefacts was a concentration of complexity in single artefacts, and human experience became distilled. The implications of the diversity and richness lost from that time must be considered.

From this point on, innovations are concentrated on the interfaces between cultures, and Chapter 3 considers other early developments in structure distinct from the Roman West. A tension field is set up between geometry and other externalisations of thought; proportional systems in buildings have remained with us since Plato's experiments with strings, but other concepts gain ground. Cultural ideals are visibly embedded within structural form; Justinian's church, the Hagia Sophia, standing on the pivot of Christianity, appropriates classicism to make a vast miniature of the cosmos, while the corbelled bridges of ancient China rely on an accretion of similar parts to form timeless wholes.

The fact that the buildings of the High Gothic age achieved a level of attenuation that would be too extreme for our modern checking authorities requires some explanation, and Chapter 4 examines the idealism and single-minded drive towards divine light embodied in the cathedral campaigns. The currents of history, war and trade spread technology, and handbooks proliferated. Quotidian structures, barns and boats, took on a greater complexity of construction, making high skill levels widespread. An intuitive understanding of material and structural form reached a precocious level ready for the decisive step that trammelled us into our current trajectory.

Chapter 5 describes the inception of modern engineering. Individuals returned to abstraction in order to master their universe, and despite subsequent setbacks their overweening confidence in particular methods remains with us. At this point personalities seem particularly important, for upon their contributions great swathes of subsequent thought swung or hung themselves up. Galileo Galilei (1564–1642) is the central figure, his place in the particular conditions of late medieval Italy the critical configuration that determined so much. Large-scale procurement began to regulate practice, and specialisation brought a focus to endeavours that simultaneously simplified and extended useful results. The success of the new methods attendant on their effectiveness meant that the introduction of new engineering met with relatively little resistance while other humanist ideas of similar import took far longer to take root and gain acceptance. The status of technique as poor relation to other endeavours seems to have been initiated about this time.

The so-called Enlightenment is considered in Chapter 6, and the mixing of consciousnesses, ancient and modern, then taking place offers insights into both strands of thought. It is this confrontation between old and new that is so fascinating, and where these parallel discourses were allowed to touch and intermingle still affects our contemporary outlook. It is not just the vestiges of 'intuitive' understanding that operate in modern engineering; a deep pragmatism still reverberates through the industry. A concern for appearances creeps in and macro-economic forces dominate innovation, requiring original responses to new building problems. Colonialism and empire-building necessitated the remote control of construction.

The relationship of engineering to mathematics is followed through across Chapters 6 and 7. A continual see-saw of influence between the two topics is reviewed to demonstrate the crossovers intrinsic and fundamental to two quite different ways of thinking. Mathematical thinking is sufficiently specific to search for forms common to all engineering practice. The impact of some of the finest minds ever born, or rather such of those who chose to express themselves through mathematics, has left legible traces and idiosyncrasies in a superficially 'pure' subject. Engineering solutions have tended to coalesce around immediate problems, but despite their contingency they have often illuminated a way towards generalisation.

The consciousness that is modernism and engineering's place within it is the central subject of Chapter 7. This is the essential condensation of our current position and the base from which we are only just beginning to extend. Why engineering itself should be so thoroughly absorbed by a single attitude so late in its development is questioned here.

Re-diversification attended the high industrialisation of the later nineteenth century and a return to technological idealism fuelled a hundred years of endeavour and ideas. Could these feelings have been maintained? Was there some failure or an inevitability about the outcome? The influence of social and cultural currents on engineering practice is reviewed alongside the counter-flow of engineering expectations informing the *zeitgeist*.

The American Civil War brought the end of any innocence engineering might have retained. The mechanisation of destruction was completed with the design of the Springfield rifle, and the introduction of long-range artillery, the ironclad, the submarine and aerial observation detached protagonists from one another. Chapter 9 starts with the reconstruction, which consolidated a power-based form of engineering that has gone on to be a mainstay of contemporary globalisation.

Engineering rebuilt another high heroic age, in the first half of the twentieth century, reaching a second culmination in the Manhattan Project. Chapter 10 describes the steady assembly of a sensibility, and Chapter 11 its descent into the two world wars and subsequent recovery, renewal and transformation. The truly tragic outcome of the conflict and man's drive for dominion over nature, nuclear science is coupled irrevocably to our highest endeavours and has coloured everything since its inception. The historical survey finishes with speculation on what it must have been like before we had to surrender to the irreconcilable.

It is at this point that the equating of engineering with art takes a widespread if tenuous hold. The many and varied products of engineers' still largely unselfconscious efforts are recognised as artefacts of critical importance. Engineering is aestheticised and appropriated by cultural critics along with many other marginalised discourses. Does our field have something in common with disciplines such as advertising, restaurateuring, journalism?

Chapter 12 concludes by attempting a survey of contemporary trends, all the time acknowledging the absence of critical distance. The point is made that we do not try to identify developments from which future events might be predicted but instead examine a continuous present being remade. The here and now and the presence of the past within it is the object of study. Generalisations are made about this condition of sameness in the diversity of events. Adopting the definition of history as a succession of current mental states, this study emphasises the different positions of a few protagonists in respect of their activity. The changing meaning of these relationships especially reflects on our times. What we understand by engineering and its history shifts and re-centres itself.

Economies grow and materials improve, but the transformations that we are interested in as designers will now come from the choice of problems addressed and the conceptual tools applied in their solution. Second-order effects, disturbances additional to the main actions within structures, are now on the cutting edge of research; their significance is seldom determined at the outset of an investigation. Amenability to computer analysis, a subject now dominating the world's education systems, may be the determinant of these preoccupations. Does this amount to a debasement of engineering? Improvement in predictive ability makes for a certain sloppiness not yet replaced by the next rigour.

A book is reductive, and fascinating personalities, their products and projects have been excluded for want of, space. The author's choices are made in order to build an argument, with a large allowance for aesthetic preference. At the scale of our industry's group consciousness, other processes must also be at work. How are some case studies and individual contributions, celebrated structures and historic incidents, identified, standardised and canonised at the expense of others? There is as yet no established critical superstructure for engineering. In the visual arts, architecture, literature, sport, business, even advertising, there are distinct roles played by those who illustrate, interpret and shape the tireless output of artists. Good engineering is currently identified by peer approval. Modern engineering, of the kind we shall see established by the French in the eighteenth century, has traditionally been governed by a conservative establishment, and change and improvement have come from a surprisingly limited number of sources. Commercial pressure is probably the weakest of the natural forces with individualism, and the need to shape a professional career being much stronger drives. There is continual pressure to introduce new methods and understandings to gain recognition. There are still new building problems to face but no longer completely new challenges such as those the nineteenth century called forth.

An important criterion used to select one type of engineer studied here has been their relationship to their contemporary orthodoxy. Typically they will be working in a milieu where standards have been established. These may well be evolving but slowly, and the engineers focused on are part of this process but also concentrations of the group consciousness of which they form a part.

Another obvious strand includes engineers who are worth special emphasis because of their depth of thought or because significance has condensed around them. The question of individualism in engineering touches some deeply rooted preconceptions. Subrahmanyan Chandrasekhar's proposition that the Sistine Chapel ceiling required an individual but Newtonian physics only the inevitability of modern man sets up a sophisticated argument for a common creativity.[3] However, engineering sometimes seems to be an exercise in reducing or eradicating authorial voice, and many engineers celebrate examples where purity of form and abstraction preclude the art of the individual from obscuring deeper reverberations.

Emphasis on biographical detail excludes the anonymous work of the everyday. Quotidian building must be considered in a different way from prototypes and exemplars, scanned lightly and not arranged too comfortably into types. The established canon comes from elsewhere, and anonymous examples offer their own paradigms and variations from the norm. Mapping the tensions and separations between what is regarded as the conventional standard and the departures from it is an established means of determining design vigour.

Engineers and projects are given equal emphasis in these pages. Do the various relationships of engineers to their projects have any coherence at all? Studying the characters, we should also analyse the embodiment of their ideas and the extent to which they can be captured in the outcomes. Bits are filtered out or lost while other parts take on special significance, emphasised by the nature of the medium. Nature and society modify the most precise intentions. If these influences are understood and can be analysed then ideas and their interactions might be further separated out and perhaps recombined in new ways – a parody in design philosophy of Baconian scientific method.

Engineers obviously do not need these sensitivities in order to function. They absorb contemporary sensibilities osmotically, through living and working in a culture. Engineering can, however, be viewed as a beautiful artefact made up of historically coherent expressions, and studied and enjoyed as such. This approach might further the development and reach of modern practice. A recent history book used structural engineering as a trope for the way the Spanish organised the colonisation of South America in the sixteenth century.[4] Various concepts from structural engineering are used to illuminate the processes of statecraft. This delightful conceit suggests in its turn the recasting of engineering using other discourses which can only enrich it in perhaps completely unforeseeable ways.

So this collection of histories sets out to become a rather 'de-constructive' handbook of engineering; the reader is invited to build an awareness of the precedents and mechanisms that condition current practice. Self-consciousness stifling creativity is a danger previously mentioned. The engineer responds by observing the moral obligation to operate only with the deepest knowledge and control if that engineer be allowed to act at all.

Chapter 1

# Prehistory and ancient times

The archaeological record is not the only place where one finds the traces of ancient builders. As technologies have proliferated and means and ends have multiplied, a handful of ideas that humans use to alter nature remain unchanged. Without calculation or computing science, structures have always reached a common level of complexity and maintained a consistent proximity to society. Good engineering transcends both time and place. In the following collection of people and projects, all those referred to would recognise one another's actions whatever their separation, and this recognition, the 'continual present' in engineering, is the subject of this book.

## Myths to organise thought

Older conventions have special importance, their invisibility seeming to be in direct proportion to their ubiquity. Ancient societies made up creation myths – 'back stories' serving to impose coherence, sense and value upon their present. They usually involved the generation of the universe, cosmogony, followed by the maintenance of a status quo. Some higher power knows how to order things, and it became accepted that things could be ordered. The will to survive transmutes into a desire to understand the environment and a separate but parallel urge to control it. Human consciousness seems to have evolved out of a two-way process, responding to its surroundings, creating and reflecting social patterns, building on subconscious foundations, making the world around itself as it was itself being shaped.

Our physiologies reflect the influence of our development of tools. And perhaps our minds are made to match those particular mental processes with which we have chosen to apprehend and manipulate the world's potential. Man's ascendancy is based on control, social control and the conquest of nature: politics for the first, technology for the second. These two aspects of civilisation seem inextricable. The ancient Greek word *cosmos* stems from a collective noun for a group of men, the minimum needed for a task, later coming to represent an abstract unit. It immediately sets up a frame of reference, a limitation on action. The first thing

1.0 *(opposite)*
**The collapsed pyramid at Menium, Fayoum Oasis, 3000 BC**
The failure of this prototype, despite its sophisticated differentiation of core and covering, forced Pharaoh Sneferu's builders to rapidly improve their expertise. Within one generation they were able to move on from low mastaba tombs to creating the Great Pyramid of Khufu.

an engineer always does is delimit his sphere of operation and his consequent responsibilities.

## Technological apartheid

The earliest times have been subdivided according to their technologies, particularly the materials used: stone, bronze, iron.[1] The time spans of prehistory, however, produced an apartheid, with different stages of development co-existing, and these anomalies challenge the notion of a neat progression. All the effort of these times was to conserve; thus innovation acquired its subversive component. Bronze was little sharper than flint and developed to support social display. This pre-adaptation as ornament, in ceremonial swords and shields, then permitted metallurgy to evolve practical metal weaponry coincident with the urge towards the first militarism.

## Prehistoric technology

The late Neolithic of northern Europe preserved some stamp of prehistoric societies as urbanism meanwhile slowly took over in Bronze Age Mesopotamia and ancient Egypt, imposing restrictive infrastructure and organisational requirements. Alexander Thom's book *Megalithic Lunar Observatories* opens with the comment: 'Ancient man's geometric knowledge will never be fathomed but that he was an engineer is not in doubt.'[2] The author's conclusions are drawn from painstaking surveys of Hebridean stone rings, which he sees as practical tools for safe navigation. Their builders had reached a fundamental conclusion about the repetitiveness of nature. Permanence resided in the heavens; the moon rises in a pattern, and that pattern can be abstracted and recorded in a scattering of stones. They modelled a nature concealed, making observations of one thing to predict the behaviour of another. In navigating the tidal races of the islands, they had coupled the lunar phases to the movement of the water. The first engineering used physical measurements to overcome an inability to measure time. The notion of linking measurements of space and time is even older, dating to the domestication of the horse, a day's walk differing in distance from a day's ride.

Other researchers have found geometric concepts embedded within Neolithic sites – proportional systems such as the Golden Section, which made the proposal of a standard unit unnecessary.[3] Thom's point about geometry is, however, well made. Once the method of siting using foresight and backsight had been hit upon, the layouts were self-generating. In examining artefacts ten times older than the use of writing, it is impossible to be sure whether an ellipse or any other abstract form had indeed been recognised at all, but a way of using physical space had been established.

Many influences on Neolithic building technology can be inferred from the rings (**figure 1.1**). Their form recalls the simplest henge, the circular space defined by the maximum clearing for a given effort, a timber then stone enclosure realising concepts of separation and then reconnection with a larger universe. The component

**1.1**

**The Bluestones, Avebury Rings, 2000 BC**

These stones from the Welsh hills travelled 100 kilometres by sea, river and overland. Weighing 30 tonnes each, they have a minimum of squaring and are embedded just enough, given ground conditions, to survive thousand-year storm events. The builders kept to a minimum of means.

stones reflect their geology: 'doggers' or loose stones eroded from a base, erratic boulders carried by ice. The limits of workability, bruising softer stone with harder to give shape, enforce a harsh hierarchy of finishing quality. The spigot connection between the uprights and lintels of Stonehenge, beaten from the parent stone instead of formed with an inserted key, are an astonishing profligacy of effort for an invisible and seemingly irrelevant improvement to the joints. Overall scale seems to have been capped by transportability: that is, determined by the manual effort that could be concentrated to manoeuvre dolmens on log rollers and mud slides, perhaps a team of twenty, pulling a weight of about five or six tonnes – a modern-day truckload – without significant inefficiency.

Primitiveness only resides in the means that ancient man had to hand; the development and refinement of the available technologies was complete. Stone tools, mined in a centralised industry, pre-finished and then distributed across a vast trading area of Europe and Scandinavia, were worked up into beautiful implements with an anthropomorphic quality lost to modern production.

## Building in ancient Egypt

Ancient Egypt's social structure developed around the exploitation of the Nile's flood cycle, and agriculture required large-scale organisation to exploit the annual renewal. Out of the abundance came a centralised economy, with population concentrated in the river valley, and from this conformity an army could be drawn to defend and expand the empire. Such an institution in turn sought monuments, an iconography of permanence to bolster its prospects, and the period of inundation allowed the entire population, otherwise idle, to be mobilised in building projects.

Their creation myth tapped directly into experience: the rising of a primeval hill from the waters. Joyce Tyldesley charts how the forms of simple funeral mounds developed into stone mastabas, then stepped pyramids, finally becoming geometric prisms to meet eternity.[4]

## Pyramids diversity in their construction

The construction of the Old Kingdom Pyramids has been the subject of wide-ranging speculation shading off into the mysticism of 'pyramidology'. Two notions worth examining are firstly, that their construction techniques evolved only very gradually in surroundings of almost immutable conservatism, hidebound by tradition; and second, that they were carried out at an almost superhuman scale and accuracy.

In fact the static systems needed to raise man-made mountains with deeply embedded artificial caves was worked out across only three generations.[5] The archaeological record includes, however, enough failures to show that structural experimentation was continuous and diverse. The large number of unfinished projects, some with brick cores, mortar, sand fills, corner dovetails and other innovations, and a succession of mistakes and failures littering the desert reflect the frantic pursuit of refinement and economy as the kingdoms came under pressure in their respective declines.

The Pharaoh Djoser's stepped pyramid at Saqquara is held up as the transitional example between mound and prism, and successively smaller versions lie embedded within the finished pile. The architect Imhotep extended his startling originality in pyramid design to other equally innovative elements of the surrounding complex, but precisely why there was such an abrupt departure from tradition remains shrouded. Written references to building works per se are sparse. Parkinson notes that urbanism and technology were not reflected in the ancient Egyptian literary tradition because at the time they were recognised as complete and parallel discourses in their own right,[6] but it increasingly appears that cross currents from political, theological and even agricultural development interacted strongly with the campaigns of monumental building.

The structure of the stepped pyramid comprises sloping leaves of masonry laid against a core, like the layers of an onion.[7] This is not ideal. A modern engineer analysing the likelihood of the slope's collapse might imagine virtual slip surfaces in such locations when trying to work out the balance of stability, so introducing planes of weakness isn't helpful. On the other hand, the settlement of the whole mass on its plinth of rock or sand can be much better controlled during construction, and each layer, being a small proportion of the whole, can be realigned as work proceeds. Practical considerations took precedent over structural effectiveness.

Graffiti indicate that Imhotep went on to serve the next pharaoh, but two subsequent stepped projects went unfinished before Pharaoh Snefru appeared to preside over some hard-fought-for departures from convention. He inherited a period of prosperity and commenced a stepped pyramid at Meidum, but after fifteen years the site was abandoned and a true pyramid was started nearer to his capital. The silty ground at the new location was a poor foundation, and improved quarrying, bringing bigger blocks, faster aggravated settlements. The work too advanced to start again, the builders instead adjusted their angle of slope and the result was the bent pyramid of Dashur. In their building work the ancient Egyptians demonstrate an admirable ability to cope and adjust to new opportunities and setbacks.

Back at Meidum, Snefru's first project was reopened and the tiered structure cased in smooth outer planes, but the bonding was inadequate and some time after the construction a long sequence of failures commenced, with shedding back down to the monument's central core.

Wisely, and perhaps surprisingly, the authors of these debacles were not destroyed by their master but were instead allowed to go on building. In the next project, for Snefru's son Khufu, one of the greatest monuments of humanity came together in a form seemingly fit to last forever.

## Construction management influencing form

Well-developed logistical tools and man-management systems made it practical to assemble the three great pyramids (Cheops' pyramid contains 20,000 tonnes of masonry) on Giza plateau. Each was completed within less than twenty years, a programme that met their function as dynastic tombs. They were just as big as economic output allowed. As to their celebrated accuracy, they are no more perfect than the methods developed for land surveying, levelling for irrigation and setting-out for taxing the harvest permitted, and they are entirely consistent with the skilled use of simple water levels, ranging rods and measuring chains.[8] Errors carried through a project or corrected during progress can be found in the archaeology, and it might be instructive to compare their frequency with those found in modern practice.

The Old Kingdom monuments have a curiously modern feel and their scale couples them to latter-day civil engineering solutions. However inappropriate it may be, it is impossible not to back-engineer our own concepts into the mindsets of the ancients so alien to us now. Simultaneously, some things unfathomable may be carried forward to us – conventions of construction and excavation that we maintain but that relate only to some requirement long since lost. Their resolution of the problems of stability must have stemmed from the close observations of slope failures, familiar from the bank collapses in the canal systems and the shifting sands of the Nile delta. They grasped patterns and projected them onto the unknown conditions that they wished to control and their built-up multi-cored forms weathered with veneers of polished stone have lasted millennia.

As the Egyptian engineers went about the overall structuring of the pyramids, they also wrestled with the detailed problems of deeply buried chambers.[9] Vertical and horizontal pressures had to be channelled around compartments and

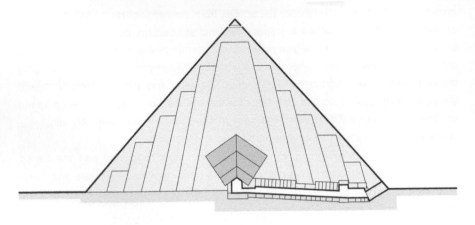

1.2
**Sahure's Pyramid, Abusir, 2480 BC**
A variety of structural devices were employed to roof over the chambers within the pyramids. Rudimentary voussoirs were superimposed and packings of aggregates used to introduce ductility between the brittle stones.

Speculative reconstructions like the one adjacent by the German Egyptologist Ludwig Borchardt (1864–1938) tried to reflect an ancient consciousness but have seldom proved accurate.

passages. Rudimentary corbelled roof systems were elaborated with primitive relieving arches, sometimes superimposed into sets (**figure 1.2**). In these design exercises, load paths were having to be envisaged. Gravel packing between the layers dispersed pressures across the brittle stones, a device gleaned from lifting and moving processes.

## Space conception contrary to technical capability

Besides pyramids and tombs, the Egyptians have left us with the remains of temple complexes and irrigation works. The loss of more prosaic building makes it difficult to assess the overall technical competence or outlook of the time. The repetitive sequences of building and additive layouts do emphasise concepts of continuity, an endless *ever-present*. The architectural theorist Sigfried Giedion found in the temple buildings an attitude towards space different from our own.[10] The crowded hypostyle halls do not fully exploit the technology then available. The columns could be much more widely spaced. This under-utilisation of structural resource he assigns to a rejection of infinitely extendable space, a consciousness strongly contrasting with our own, but it seems simpler to see the closely spaced columns as part of a process of capturing and re-representing natural forms (**figure 1.3**). Timber columns, then stone shafts, were shaped as lotus stems for religious signification and closely packed to resemble fantastic papyrus groves – symbols of peace and plenitude to the Nile-dweller.

Masonry construction ensured permanence. The classical historian Pliny describes the ancient Egyptians quarrying and moving their material.[11] The variegated geology of the valley made various types of stone available from near and far, and the annual flooding allowed very large components to be prepared on barges in the dry, then moved economically over large distances. The boats used to carry the largest single stones, obelisks, relied on technologies of timber and hemp. Water transport initially developed around reed boats trussed and shaped from tight-packed bundles (**figure 1.4**) and these forms were subsequently transferred to timber construction, as butted planks laced together. A relief from the temple of Queen Hatshepsut

1.3

**Temple of Isis at Philae, Nubia, 280 BC.**
**Watercolour by David Roberts (1796–1864)**
Temple columns were embodiments
of the Nile reed beds and hunting grounds,
the ancient Egyptians' idea of paradise on
earth. There was no incentive to abstract the
columns, made by bruising down the stone
with diorite, to minimal cylindrical shafts.
The over-packing of the entrance halls
conjures a very specific spatial effect.

(1473–1458 BC) in Deir el-Bahari depicts a massive timber hull strengthened by cross frames and pre-stressed with rope girdles. Inverted trussing, adjustable by turn-buckles, carries the weight of prow and stern towards the centre of buoyancy. The cargo of the barge depicted comprises a pair of obelisks, their weight difficult to determine but likely to be in excess of 500 tonnes each.[12]

## Greek thought and construction

Elsewhere in the Bronze Age Mediterranean, other cornerstones of western science and technology were being laid. The pre-Socratic philosophers of ancient Greece had worked and thought through times as Stone Age culture turned to Bronze, a

1.4

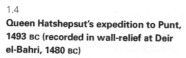

**Queen Hatshepsut's expedition to Punt,**
**1493 BC (recorded in wall-relief at Deir**
**el-Bahri, 1480 BC)**
Five ships journeyed in search of cedar wood
towards present-day Somalia. The event is
recorded in fine technical detail, showing hull
bindings bow and stern, an adjustable hogging
cable keeping the planks together and thwart
ends appearing through the hull. Crew men sit
in the sail to stop it backing against the light
hot airs.

transition marked less by revolutionary innovation than by cultural cross connections and diversification.[13] The philosopher of science Karl Popper describes an age of 'enlightenment', with thinkers coming to grips with what the world was like and what could be done with it.[14] The pre-Socratics released man to acquire knowledge, their separation of body and nature a prerequisite for reflective thought and engineering ambition.

In these ancient debates widely disparate concepts of the universe followed on from one another. On the one hand, the world was thought to be full and therefore unchangeable; alternatively, it was completely empty and undetectable. It might be totally mechanistic, programmed onto an inevitable path, or else everything was intractable chaos. Parmenides (b. 510 BC) recognised a world in which absolute knowledge is embedded and concealed and where, although everything is explicable, paradoxes nevertheless multiply in the face of our efforts to understand. He found permanence only in geometry; the abstractions somehow got behind the appearance of things. Accepting the existence of a coherent system out there and our own understanding as but a poor substitute for knowledge, human effort becomes directed into developing inquiries, ways of searching, *meta-hodas* – methods – to cross this divide. Engineering is one such method.

The idea of *anamnesis* was central to this thinking: that knowledge is discovered, or rather recovered. It incorporates the further notion that the soul has previously existed in a purer state and therefore gained a catalogue of unadulterated ideas, which accounts for the continual resurfacing of ways of doing that this book seeks to portray. Knowledge has not developed by accretion, but instead we have only better approximations to the same truths. For *techne*, the practical art of making, these concepts allow any number of approaches to a task to be equally valid, subject only to their coherence in some larger totality.

A sequence of thinkers has contributed to the assumptions we retain and work to. Plato (429–347 BC) asserted that underlying patterns and forms – the invariants that make engineering possible at all – can be retrieved from the apparent chaos of nature.[15] In the *Timaeus* he posits that everything reduces to triangles to explain the irrational, in a weird premonition of modern computer analysis.[16] Aristotle (384–322 BC) contended that we should rely only on our immediate sense experience – the grounding for a good engineer's continued reliance on close observation. The Greek aesthetic idea of balancing intellectual and physical worlds keeps a hold on modern design.[17] Socrates (469–399 BC) centred his work on a persistent theme of critical analysis, with ideas to be held up, examined and super-seded in a continuous process.[18] We now take it for granted that our understanding continually moves on, enriching itself in experience.

## Early mathematics

The development of mathematics was inseparable from these enquiries, and Serafina Cuomo draws out two strands, residual to this day.[19] First, Greek ideals of democracy and public accountability linked to programmes of colonisation spurred on advances

1.5
**Doric Temple,
Fruli, Sicily, 350 BC**
The sacred grove
ossified. The early
colonists built their
shrines of bundled
timber, lathe ties,
beams and pegs.
The elements were
gradually replaced
by the permanence
of masonry. The
surface forms of
the original material
are retained over a
substrate
determined by the
workability and
transportation of
the stone.

in practical surveying and accounting. Land was to be divided fairly and public money seen to be disbursed appropriately, and a general numeracy was established. Second, and simultaneously, the philosophers' detection of divine depths in geometry and the practical results of its application – Archimedes' lever and so on – promoted the search for mathematical axioms and ever more powerful methods.[20] The irrationality of number theory had been recognised and geometry was preferred; the beauty of Euclid's elements continues to be celebrated.[21] Engineering and mathematics were to be inextricably linked from then on, and the bonds would be quantification and abstraction. A project became an economically definable task and was seen as a particular case of a more general potential.

Despite these discoveries, the logical foundation of engineering was not established sufficiently for everyone. After centuries of practical outcome, doubts over the validity of concepts such as force vexed Ludwig Wittgenstein (1900–1960) to the point where he abandoned structural engineering to study the foundations of mathematics.[22] His findings on the limits of logic have not yet affected engineering mathematics.

Greek temples of the Classical age embody the philosopher's sense of an ideal lying behind a turbulent world.[23] Artists concentrated on perfection of form immediately after the threat to Greece from the East had been overcome. The temple architects created simple structures, logical, complete and hermetic, the outcome of rigorous codings (**figure 1.5**): thus from a fragment of a fluting retrieved from rubble fill, the entirety of a lost building can be recreated. This proportioning and the contiguity of part to whole represent a conception of the universe; that these perfect forms were then visually corrected, columns distorted to appear straight, and wide staircases cambered to improve appearance as far back as Minoan times implies a divinity on the very edge of real existence, outside of nature but within our perception.

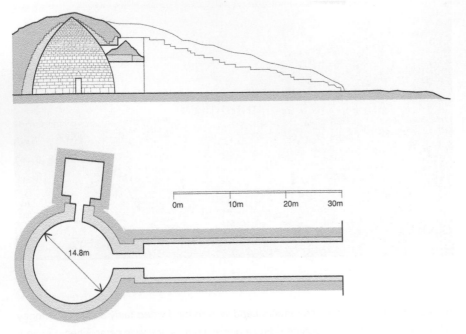

1.6
**The Treasury of Atrius, Mycenae, 2000 BC**
Pseudo-domes, comprising superimposed rings of elements corbelling inwards, appear independently worldwide. Igloos and the granaries are examples. The simple bread ovens of the early Greeks were appropriated for cremation; constructed larger, they held the belongings of the deceased before finally becoming the huge and impressive stores of Royal wealth.

## Pre-adaptation in early Greek technology

The temples were timber huts recreated in stone, their pegs and beam ends ossified into decoration. These 'skewomorphs', forms suited to one material appearing in another, seem initially to have been the outcome of processes of progressive replacement. The travelogue of the Greek geographer Pausanius records one shaft of oak remaining in the Temple of Hera, which had been converted from wood to stone over a 700-year period.[24] The influence that material properties have on form and proportion can be heavily veiled by such processes.

Structural form developed through pre-adaptation recurs in Greek technology where a system developed for one use would be appropriated to another. Relatively sophisticated technologies developed for public buildings might be downgraded to more utilitarian and everyday construction. The beehive tombs of the Minoans were the wonder of later Greeks (**figure 1.6**), and in attributing such structures to mythical beings Pausanius illustrates that construction knowledge was being lost as well as found over time.[25] The corbelled domes of simple bread ovens had been expanded to make cremation chambers and then, traditions mutating, burial chambers formed as true pseudo-domes. The simplicity of these enclosures, each ring of stone set without temporary support, atrophied the development of more sophisticated vault types.

## Eupalinos and Tridon

The spread of Greek culture, through trading and colonial expansion, westwards and into the Black Sea relied on the technical development of ships. Fernand Braudel

**1.7**
**Athens trireme reconstruction, 500 BC–AD 1978**
Archaeological reconstruction relies on morphology, how well an artefact fits its environment, or rather how accurately that environment can be defined. No example of a Bronze Age galley has been recovered, nor can the original designer's conceptual space be recreated, but the warship's specificity of purpose means that much can be determined.

describes the difficulties of communication and agriculture along the stony lands and steep eastern shorelines of the Mediterranean;[26] the seaways were far better routes to control. For two thousand years, oared warships – triremes – became the weapon of choice for the thassalocracies dominating the region.[27] Whole forests disappeared in the making of navies, and control of a dwindling timber supply enabled the Macedonian king Philip (360–336 BC) to have firstly a choice of favourable alliances and then the capability to initiate the conquests that were completed by his son Alexander (356–326 BC).

Although no remains of an actual trireme are likely to be found, much work has been done reconstructing their form (**figure 1.7**).[28] Continual fighting honed vessel design, engineering hulls, rigging and weapons to an unprecedented pitch of refinement, and all the construction was directed towards extreme lightness. A sensibility for timber construction was established, with maximum robustness for a minimal use of material. Among the smaller city states and colonies, manpower was always in short supply. Crews were skilled, trained volunteers rather than emaciated slaves. Long hulls built for speed were pre-stressed and locally reinforced to sustain the shock of ramming the foe. A modern reconstruction has been completed making full use of contemporary knowledge, and performance obtained closely matches the literary record. Despite the millennia separating them, it appears that modern designers will not surpass the original creators,[29] for such systems could be perfected within their own frames of reference using only the old methods.

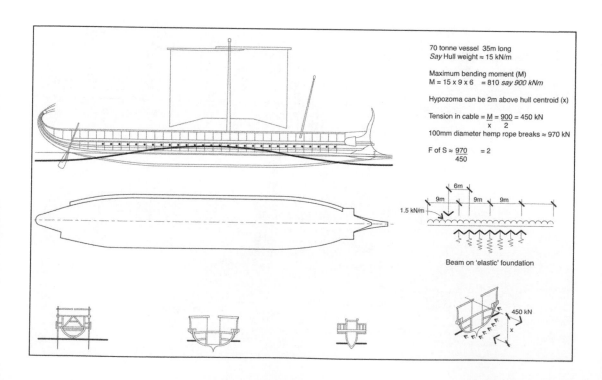

70 tonne vessel 35m long
*Say* Hull weight ≈ 15 kN/m

Maximum bending moment (M)
M = 15 x 9 x 6   = 810 *say 900 kNm*

Hypozoma can be 2m above hull centroid (x)

Tension in cable = $\frac{M}{x}$ = $\frac{900}{2}$ = 450 kN
100mm diameter hemp rope breaks ≈ 970 kN

F of S ≈ $\frac{970}{450}$   = 2

1.5 kN/m

6m
9m   9m   9m

Beam on 'elastic' foundation

450 kN

## Engineers and social function

The technologies of ancient Egypt and Greece developed out of very different circumstances and took on different social functions. That of the Egyptians seems to have appeared, apparently fully formed and immutable, after a protracted period of assembly, a kind of condensation rather than linear progression. We have seen the speed of change when it came. One might expect these processes to have been over-determined and therefore unstable – revolutionary, a long build-up before a very short period of realisation – and so it seems to have been. After each successive transformation, the Egyptian way would be fixed and would then resist change. The evidence matches Marxist theory  that technological change depends on social change. Fixed technology permeated their society, with the Nilometer, a device measuring the level of the annual inundation, used directly to set tax levels on the future harvest, with no regard for development.

## Hellenism

Why did Greek technology appear instead as an inexorable advance, continuous and ascendant? The infighting of the states promoted innovation and the compactness of the social systems permitted specialisation, while experimentation could be undertaken in the relatively benign conditions of the eastern Mediterranean. Hellenism brought a flexible urbanism and infrastructure development along with it, which could absorb alien ideas.

1.8
**Pharos,
Alexandria, 285 BC**
Alexander conquered the known world. Hellenism was supported by Egyptian grain exports. The lighthouse marking the Nile entrepôt of Alexandria, an important navigational aid in difficult waters, was exploited as a spectacular demonstration of technical prowess and power.

When Alexander came to Egypt in 323 BC, the year of his death, he was hailed as its liberator from the Persians. Under the Ptolemies, Alexandria became a showcase of Greek culture, a beautiful city with the great library a centre of Greek learning and the Pharos parading a conspicuous display of technology in front of the native Egyptians (**figure 1.8**). The lighthouse also provided a practical solution to the treacherous harbour approaches. A technical excessiveness might be detected in its height, topped by a continually burning fire, fuel wagons on spiral ramps, a camera obscura spyglass and Archimedean burning lens, a death ray to destroy ships; but the Pharos at Alexandria was a signal to the wide world. This use of technology persists to this day and is characteristically western. It was transferred through the Roman appropriation of Greek culture.

Chapter 2

# Rome and the East (220 BC–AD 533)

## Greece and Rome

Victory in the Punic Wars forced the ancient Romans to overrun the eastern Mediterranean and in turn led to the collapse of their Republic.[1] They assimilated Hellenistic culture in complicated ways – not so much by the appropriation of alien art, science and political systems into their militaristic society, but more through the use of Greek forms as sites where hidebound Roman traditions could be experimented with, tested and melded into something new. Augustus used everything Greek to reshape government and establish empire.[2] The way in which Greek technology and precepts for engaging with the real world, *praxis*, was grafted onto Roman regimentation somehow established a division that has left us with a persistent fault line between theory and action. The subsequent history of technology is one of flamboyancy reined in by pragmatism; intellectual ambition and individuality confronting custom and practice.

2.0 *(opposite)*
**Puente de Alcantara, Tagus River, AD 100**
Simple semi-circular arches of dry-jointed granite voussoirs symbolise the permanence and stability of the Pax Romana. The elementary form is easy to encode, set out and construct. One arch of this bridge was dynamited during the Spanish civil war but the remainder proved robust enough to accept a modern replacement.

## Roman reliance on problem-solving

Certainly the city of Rome, its army and then its empire were always short of manpower, and compensation was found in organisation and technology. As Roman expansion accelerated over the two centuries crossing into the first millennium, so too, fundamentally, did its use of, attitudes towards and expectations of engineering change. The Imperial age raised new concerns for the engineer to address as well as re-interpretations of accepted ideas. No longer just required to conserve a status quo, technology had to be communicated over huge distances and adjusted to new locations. Construction carried within itself new meaning. Julius Caesar's (100–44 BC) incessant activity established an outlook that saw life as a continual problem-solving activity in which the solutions would be technical. The Augustan idea that Rome was not just to dominate the world but *was* in fact the World itself meant that their technologies would have to be interior to this monolith and complete within themselves. These changes proved to be more than the veneering of one outlook upon another and also involved an underpinning of conceptual thought to apologise for a system. This embedment gives particular precepts a persistence that shapes our actions today and distracts us towards material that we continually rework at the expense of newer ideas.

## Technical development and imperialism

Roman engineering encodes the imperial problem: the reconciliation of eternal expansion and essential consolidation. The army started out as a levy of farmers, yeomen perhaps, but they knew how to dig, and during the Republic's ascent the army exploited temporary structures and fortifications to the full. As well as the standardised camp suited to a compact army in foreign territory, they also deployed man-made obstacles and circumvallations on the battlefield itself to overcome the odds. These systems had to be rapidly demountable; for example, crossings could quickly pass from attack to defence. The reversibility of the Sublician bridge over the Tiber was protected by religion,[3] and it remained a timber openwork structure, loose-cramped with timber pins, no iron, long after other stone-built crossings had made it an anachronism.

## Engineering as propaganda

The example of Sulla the Fortunate (138–78 BC) educated Julius Caesar in his dealings with the Republic. Power required reputation, and spectacular success was essential to match the conquests of his rival, Pompey the Great (106–48 BC). As his career progressed, Caesar's grasp and application of military engineering changed emphasis as textbook fortifications and siege works gave way to ever more showy constructions. A centrepiece of his self-initiated campaign in Gaul was the erection of a bridge across the Rhine near modern day Koblenz in the campaign season of 55 BC (**figure 2.1**). It represented the first step in an extremely risky raid of little strategic significance; to use the expedient of a pontoon bridge would have been to go by the book, but instead a trestled roadway was installed on piles. As we shall see, this structure acquired iconic significance for later generations, but at the time it was carried through as non-standard army procedure at an unprecedented and awe-inspiring scale. Caesar's underlying problem was political, and he wanted to impress both the German tribes and an audience back home. His wars were unsanctioned adventuring with no clear military objective, and his resources were inadequate to control and subjugate the conquered tribes, which meant that his campaigns had become repetitive attempts to quell rebellion. The bridge across the Rhine was conceived specifically as a demonstration of power, a structure capable of dominating the river itself – a potent symbol of Celtic/Germanic animism. The message was of superhuman power and permanence. Wisely, the expedition was withdrawn and the structure demounted before the spring floods came.

The sheer scale of the undertaking had conscious overtones of Alexander's massive construction effort at the siege of Tyre (332 BC), in which a seemingly impregnable island citadel had been taken by an artificial isthmus built with huge effort. The symmetry of the two endeavours would have appealed to the well-read Roman general.

**2.1**
**Julius Caesar's**
**Rhine crossing,**
**50 BC**
The pressure of the
current made the
cross frames
distort towards the
downstream.
The scissors closed
onto the cross
logs and bound
the joints tighter.
The assembly
progressively
stiffens up.

In his propaganda piece reporting his achievements, Caesar makes much of his personal involvement in the design of the bridge,[4] as practical large-scale engineering was recognised as a noble and difficult task. In his report, however, he misses one of the main subtleties of the braced framing: that the pressure of water current on the diagonals locks the system ever tighter, suggesting the plagiarism of more expert advice.

## Military display transmuted into public works

By the time of his last return to Rome, this engineering animus of Caesar's had transferred itself to a public works programme spending the incredible personal wealth that conquest had produced. The polis were delighted and overawed. Building could be a tool of political propaganda. For an ancient Roman, a Triumph was an occasion of the greatest religious significance, a celebration of Fortune granting victory. Only engineering achievements, roads and water supply were celebrated in the same way.[5]

## Engineering reinforcing power structures

Caesar was assassinated, but the Republic still came to an end. Augustus and then Tiberius made their assumption of power more palatable by emphasising public service; games and the corn dole were not the only palliatives. The Claudians adjusted the ideal of engineering towards the provision of infrastructure as a public duty, with political and social stability embodied in stone by a beneficent state.

Augustus could afford it from the spoils of his victories in the East, and keeping up with such largesse occupied potentially hostile senators. The Imperial family continued to consolidate their power base, only implicit by social acceptance, through public works in which a certain kind of large-scale but unostentatious project was favoured. Julius Caesar had celebrated in the traditional manner, by building marble temples and triumphal arches. August and Tiberius were to stamp their authority in a subtler way, with practical works executed without decoration, retaining some of the brusqueness of army command. Agrippa Postumus (12 BC–AD 14), Augustus's admiral at Actium, became indispensable in peacetime as the emperor's drains and aqueducts man and, however uncouth, contrived to marry first his niece Marcella and then his daughter Julia. The Cloaca Maxima, the origin of modern sewerage, was a source of extreme pride to the city districts through which it passed.[6]

Augustus presented his innovations as recoveries or nuances of established traditions, which meant that, in building, structural systems, architectural orders and decorations always looked conventional. Behind the traditional façades, however, ingenuity and innovation kept up their pressure. The composite order, columns and entablature, made its appearance, taking many years to find acceptance but never quite subsumed either.[7] Despite all the riches, continual expansion and a vast increase in scale brought economic pressures to bear on the construction process. Stone sheathings replaced solid masonry and the huge vaults and plinths of public buildings were framed with complicated configurations of relieving arches.

The empire's only real stability as it expanded seems to have been a dynamic one. Imperial buildings were constructed of ever more exotic materials imported from the new conquests. The archaic ritual of *evocatio*, calling on the enemy's gods to change sides before a battle, was alluded to through representations of their material surroundings, and new temples were dedicated to successful campaigns and recent acquisitions. In their fabric they used marbles, stones and timbers taken in tribute, displaying a microcosm of the expanding empire for the illiterate mob to see and comprehend. This conspicuous consumption of fine materials, teak and figured marbles, reappeared in the shipping offices of Victorian England.

## Assimilating otherness

Such acquisitiveness extended beyond building materials. The ancient Romans were disciplined but pragmatic and adaptable; if foreign ways were better, they would adopt them. Differing methodologies could co-exist, and techniques often needed adjusting to meet the widely different environments that the empire embraced. The Mediterranean fleets were burnished to flare fearsomely while British Channel ships and their crews sported a uniform flat grey to exploit surroundings where camouflage works.[8] Romans readily used technology transfers, and the *velarium*, a fabric roof sunshade over the Coliseum, was rigged by sequestered sailors.

As this patchwork of means was built up, it was simultaneously subjected to a process of standardisation. Systems of warfare, transport, building types and construction were rendered into a tradition of best practice. This focus on assembling a single value set reduced diversity in the Roman world but promoted a different kind of evolutionary process: rules give rise to exceptions.

## Method and codification: Vitruvius Pollio

The Pax Romanum of the Augustan age needed technologies to be codified and disseminated as widely as possible. Patterned like military order books and tactical manuals came an outpouring of self-help guides and standard recipes for virtually everything, from agriculture to poetic form. The writings of Vitruvius Pollio (70–25 BC) offer a survey of construction knowledge organised by a military mind, and his is the only surviving example of the many handbooks of the time.[9] This form of literature was effectively the first prose in Latin so its structure may have influenced our entire prose literary tradition.[10]

## Value systems applied to construction

Caligula and Nero both tried to resuscitate the philhellenism of Augustus to bolster their own authority, but Caligula's massive pontoon bridge between Baie and Puteoli failed to impress and backfired in ridicule and accusations of atrocity. After the dark period of confused government, the Flavian emperor Trajan (AD 53–97) took his people back on the offensive, scoring successes in the East. Nevertheless, the structure of his bridge over the Danube still held on to the old ambivalence towards empire,[11] its solid piers signifying permanence (traces remain to this day) while its demountable superstructure acknowledges a continuing insecurity. The style in which the structure is depicted on the column celebrating the campaign, in the blocky figures of pre-Hellenistic Roman art, indicates Trajan's rejection of his predecessors' values.

Iron Age engineering changed and adapted to the huge pressures of the middle and late empire. Confrontation between superpowers and the onset of economic decline were the two most significant influences that took hold.

## Building viewed as a hermetic system

Hadrian (AD 76–138) was perhaps the first emperor to acknowledge the inevitable. Although his policy of retrenchment was unpopular, it did provided a period of stability. Cheaper building methods were perfected for use along the miles of fixed frontier. A new phase of urbanism coincided with a particular flowering of complexity in building in which interlocking spaces required intersecting vaulting, promoting setting-out skills and stereotomy, the cutting of complicated stone shapes. Agrippa's pantheon was rebuilt as a spherical space containing a complete world (**figure 2.2**), with the ceiling coffers adjusted in perspective.

The defects inherent in the Roman system were not to be resolved; the extended frontier needed too much manpower to garrison. Threats from outside grew in scale and the army had to become ever more mobile, ranging widely but capable of concentrating rapidly. Lengthening trade routes improved transport links.

2.2
**The Pantheon,
Rome, AD 150**
The huge dome
sets out around a
sphere representing
the world or
universe. Distorted
coffers correct the
optics of the space.
The familiar device
of ranked brick
reinforcing arches is
mapped into the
walls and soffit, a
concealment of
how the structure
must be working.
Edge thickenings
direct the lines of
thrust downward.

Ever wider rivers were being crossed, and within the empire bridges as removable fortifications had been replaced by stone arches as symbols of permanence. Unfortunately, this meant that raiders as well as defenders could move rapidly on the highways.

## Technical development through exchange

The squeeze came from both east and west, but in the Persian encounter Roman consciousness received enrichment in a different but maybe as important a way as in the Greek exchange. From the two world outlooks, fighting or trading but continually confronting one another, ideas – structural principles included – cross-fertilised. As first the Parthians and later the Sassanids persistently reversed Roman incursions, their custom of resettling captives far behind the frontiers brought western technology deep into their dominions.[12]

## Mixing cultural motifs

The simple geometries of the Romans – rational, eminently describable and buildable – contrast strongly with the structurally more ideal forms of the Middle East. The two approaches can be seen juxtaposed in a single building, the White Palace of the Sassanid kings in Ctesiphon (**figure 2.3**), which adopts the form of the Desert Arabs' reception tent – a central arbour with display carpets hanging each side – and re-writes it in monumental masonry.[13] The replacement of mud blocks with burnt bricks in the western part of the two rivers' delta relieved material limitations on scale.[14]

The vast vault of the throne room is flanked by screen walls decorated with applied orders. These motifs are lifted straight from the aqueduct structures and tiered enclosures of Rome, but the roof itself might come from another planet; the parabolic shape is structurally near perfect. How was it achieved?

One explanation co-opts the bent reed mud huts of the Marsh Arabs; the understanding gleaned from that small-scale vernacular was magnified to huge size.[15]

**2.3**
**The White Palace, Ctesiphon, AD 550**
The tiers of applied arches are an import brought by Roman prisoners of war. The vault shape is entirely indigenous, a much enlarged homomorph of the mud huts along the river marshes. Bent reeds plastered with drying mud creep towards a parabolic profile, and brickwork has enough plasticity to follow suit.

The palace builders may have learnt from hours spent staring at their own mud walls. Bending vertical reeds and tying them into an armature, then plastering with mud, initially fully plastic but quick-drying, naturally results in a catenary profile. As a 'homomorph' in burnt brick, the roof literally soars over the great hall at Ctesiphon.

Other structural and formal innovations originated in the East, where Sassanid architects completed the resolution of circular dome set onto square space. Their spherical and parabolic shells were supported first on squinch arches crossing each corner, then pendentives, continuous transition surfaces. At one location, the Great Palace at Firozabad, the plan rises from square to circle seamlessly. The various systems, primitive and more sophisticated, co-existed for centuries.

## The late appearance of pragmatism in engineering

The Roman Empire moved inexorably onto its back foot. Civil wars and barbarian incursions culminated in the so-called Third Century crisis,[16] and economic and organisational problems had to be addressed by real change. The boundaries of the empire needed improved fortification, and manpower was dropping, stretched and decimated by famine and plague. One tendency was towards specialisation, with ever more highly skilled individuals countering inadequate resources and lagging communications in their responses to the new circumstances.

Many of the later emperors reinforced their leadership with pomp and display. Aurelian (AD 214–275) was condemned by bluff manners and military upbringing only to solve problems head on, literally knocking the provinces into shape. He halted the rapid turnover of warring soldier emperors and introduced a technology-based defensive system, and the emphases of engineering changed again. Although wealth could still focus in ostentatious building programmes, construction generally had to be carried out in extremely straitened conditions, and cheap and rapid ways of building that needed little skill predominated. No longer considered to be only temporary or eternal, buildings were being designed at the intersection of fixed budgets and finite lifespans.

Every town, including Rome, needed walls to survive the deep raids that were becoming commonplace. The deployment of artillery, machines for throwing stones and bolts, and eastern methods of siege, particularly the undermining of walls, radically changed the layout of fortifications. Traditionally military bases had reflected only the army's internal structure, a microcosm of Rome carried into foreign parts, but rectilinear camps were no longer adequate. Now the lightly garrisoned forts would have to take any advantage the ground could offer. Bastions – strong points – were set around polygonal enclosures at regular intervals matching the range of defenders' weapons. Plan arrangements reflected circulation patterns, their rigid geometries combined with amorphous space in a very modern way. Surveying techniques were developed to enable this free planning, with scarce materials and logistical requirements carefully estimated for design options.

## Alternatives to geometric forms

These departures from the rigid symmetries and closely proportioned hierarchies of traditional Roman architecture coincided with a more fundamental shift in design sensibility. Eastern artefacts seem to have adhered more closely to natural form. The Parthian bow comprised a perfect composite of horn and wood. The Romans sought to codify this strange and effective approach. The proportional systems with which buildings and boats had been dimensioned since prehistory were subtly re-centred. Siege engines were among the first frames to be stressed to a level where structural adequacy dominated the design; either the most important part is the safest, or all the elements are given precisely equal adequacy. To make the most effective and lightweight weapon, ballistas were sized according to precise rules of thumb generated from experimentation,[17] the pivot pins defining the size of the whole machine. Weapons were decommissioned by the confiscation of such vital parts: nowadays, it would be firing pins; in those times, it was the pre-stressing girdles of ships and the axles of catapults and siege engines.

## The recognition of obsolescence

Civilisations had always adjusted their building techniques to economic necessity, but it was in the Roman Empire that building resources really began to be manipulated with a clear acknowledgement of whole-life performance. Short-term structures were thrown up, with provision for more permanent works to follow. London Bridge follows the offset alignment of a temporary timber trestle alongside which a permanent stone crossing was projected but never completed.[18]

## The influence of geographical diversity

The variety of sites encountered by Roman engineers across the empire required the exploitation of all types of indigenous materials, and construction techniques were adjusted to meet as-found characteristics, exploiting strengths and suppressing weaknesses. Rapid weathering revealed defects more quickly and encouraged innovation and evolutionary development.

Materials modification is one of the oldest subsidiary tasks of construction. Ancient man worked within a found world, selectively refining organic products such as timber, leather and bone to his use. And the early metalworkers always worked in an alchemic intimacy with nature. Fire-hardening of timber and then clay presaged the Egyptian discovery and development of glass, the slaking of lime to make an adhesive mortar and later the Roman achievement of concrete – pulverised volcanic earth that hydrates into artificial stone.

## Materials and morphology

The predominance of different building materials and their associated technologies at different times seem to have both reflected and affected contemporary conditions. In the Middle East brick-making developed from sun-dried to fired to glazed to colour-glazed, and the huge increase in strength and robustness attendant on these processes led to a leap in scale of construction on the western side of the Euphrates. The right types of clay and the presence of fuel coincided with the local dynasts' will to rule. Palaces of fired earth had the names of the ruler stamped on each brick buried within the layers of the building, these glazed inscriptions meaning that history could be given a real permanence. The active vulcanism of the Campagna and further south furnished the Romans with a plentiful supply of pozzolanic ash, from which to make a true concrete.[19] The material was adapted both as ubiquitous matrix for cheap rubble walls and sophisticated artificial stone for the vast cast vaults of the late empire. A new generation of buildings, huge civic spaces expressive of revolutionary social institutions, courts and baths, realised the material's potential.

## Early foundation engineering

Ever heavier structures were being erected on increasingly difficult sites. The sprawl of established cities extended into more marginal areas. With the seven hills of Rome overdeveloped, it became the turn of the nearby Palatine marshes to receive new and massive structures. The steady trend towards urbanism founded many new townships, often on land in deltas or marshland where economic advantage outweighed construction difficulties. Foundation design was being elevated from a secondary consideration, occasionally overlooked, to a governing aspect of building form. Egyptian experience had centred on controlling settlement. Their artificial mountains had subsided on weak strata, and slope stability problems beset canals and pyramids. Early Roman strategies to deal with differential settlement involved managing movement or adopting forms capable of sequential adjustment. Preloading aqueduct piers with the full weight of stone voussoirs before they were set out into arches meant that they could be set level after a pre-consolidation had taken place,[20] and the construction of diminishing arched tiers allowed each stage to be precisely levelled despite the distorted line of the one below.

Earthworks, both military and agricultural, gave the Romans an intimate understanding of the soil. Load dispersal through soft ground was understood as well as a rudimentary idea of how ground water movements affect consolidation. Well-tried ways of drainage, developed for agriculture, irrigation and land reclamation, were assimilated to ground improvement and foundation applications. Bundles of brushwood – faggots – would be used as a draining base onto which massive stone-work could be placed, and building plinths and associated drainage schemes became inordinately massive, the weight of the base sometimes exceeding that of the building above.[21]

The principle of distributing load evenly across weak strata was developed in various ways. The Enlightenment researcher Giovanni Battista Piranesi (1720–1778) recorded Imperial construction methods in his engravings. He speculated complex patterns of inverted arches underlying the major monuments (**figures 2.4, 2.5**). The symmetry he applied to the forces at work within the gravitational half-space is an astonishing engineering insight for which pictures of reflections might have provided

**2.4 and 2.5**

**Pons Fabricius, Rome, 62 BC, and engraving, 1756, by Giovanni Battista Piranesi (1720–1778)**

Expertise in fortification informed Roman foundation design. Arches were inverted to distribute point loads of bridge or vault piers evenly into the ground layers. The near bi-lateral symmetry of this design within the 'gravity half-space' that we all occupy indicates an equality of consideration for all the parts of structure.

the clue. Rusticated bases, an adaptation of fortification, were made using masonry bonds capable of arching across weak spots. Polygonal work, suited to unbedded building stones such as granite, not only minimised material and manpower wastages in dressing stones but also allowed load paths to reconfigure as the ground shifted and settled.[22]

## Piled foundations

Some prehistoric peoples had taken to living over water for defence. The pile-dwellers of the Italian and German lakes stood within the wide invasion route south of the Alps. Their use of timber baulks driven through sediments into the grip of firmer strata or onto hardpan was appropriated by Roman engineers to deal with the concentrated loads of bridge piers and abutments built over water.

The littoral plain around the northern Adriatic afforded easy incursion into the Italian peninsula. Cities enriched by the nearby saltpans were vulnerable. As breakdowns increased, the early Venetians sought refuge in the offshore haze.[23] They began by stabilising the shifting banks of their lagoon with log palisades, and by the onset of the barbarian invasions – Alaric in 410 and Attila fifty years later – this experience had become a sophisticated appreciation of piled foundation behaviour. Inert and rot-proof in the anaerobic mud, preloaded with masonry cappings, short stubs of timber from the Dalmatian forests became artificial islands on which a fantastic city took shape. The problem of water supply was simultaneously solved by sophisticated ground engineering, and public squares doubled as rain catchments over reservoirs isolated from the surrounding brackishness.

## Scale changes in construction

Modern historians replace Gibbon's Decline and Fall of the Roman Empire with a metamorphosis, emphasising continuity. Barbarians didn't want to destroy the empire: they wanted to be part of it. Emperors became kings. A record of disruption distracts from an underlying conservatism, particularly in the East. The buildings of the late empire, under Diocletian and the Tetrarchy, then of the Eastern Empire under Constantine, highlight some of the forces at work. There is a huge jump in scale and complexity of organisation, with material advances and a series of formal inventions answering the requirements of the new public buildings, baths and people palaces. The ritual of the bath compartmentalised a sequence of experiences in which the final relaxing caldarium hall, cross-vaulted and beautifully lit, seemed to project a profound sense of permanence and repose in troubled times. These buildings offered the possibility of infinitely extendable space.

## Matter becomes inanimate

Diocletian (AD 236–305) unsuccessfully persecuted Christians as part of a wider effort to recover people's commitment to the state. Constantine (AD 271–337) opted instead to replace the pagan past and thereby secure his position as universal emperor. His appropriation of Christianity, an ideology that had hardly had time to define itself fully since its inception, and the state's subsequent immersion in the religion, is fundamental to our contemporary western outlook.[24] The adoption of Christianity and the spread of the other monotheisms dispelled, or rather de-centred, the more Roman concept of animism. Matter could be studied in itself; if a thing was cracked, this followed from an immediate cause. The operation of divine process was at such a remove, uncontrollable and unforeseeable, as no longer to be of immediate concern.

The cult that Constantine chose to promote was profoundly influenced by pagan and classical thought, both in terms of ideas and in the way it moulded itself to assimilate earlier cultural structures, forms and expressions. The Roman basilica, an aisled hall with a simple timber roof, became the standard shelter for its assemblies. The forms of the late empire were effectively frozen as the received way of church-building for all time, simultaneously shaping and adapting to ritual. Churches were microcosms of the universe. Building structure, the church and the world interacted in a fraught and eventually unsuccessful attempt at achieving theological consensus.

## Symmetry and sub-systems permitting scale increases

The mausolea of the last pagan emperors, among them Galerius (AD 250–311) and Diocletian, are centralised forms, epitomising stability and completeness.[25] The characteristics of domes and their associated support structures, ambulatories and buttressing half-domes were worked out and refined at ever increasing scale. The symmetries and hierarchical form of these propped piles permitted bigger spans and much more height, but the patterns of forces being conjured up were not well understood, and recurrent structural damage and running repairs were common.

Constantine's new religion moulded these building types to meet its own requirements. Baptisteries, Christian tombs and the chapter houses of the monastic movements appropriated centralised forms as emblems of the spiritual gravitation. The trajectory along which the Sassanids developed the pendentive linking dome and octagonal plan seems to have been followed all over again in early church buildings. Tiered squinch arches prefigured fully developed pendentives, technical devices that seemed to attract complex decoration. Classical sources were not rejected but their influence toned down to subtler influences, and the prescriptions of Vitruvius remained available to builders. Various Christian ground plans incorporate ratios and number sequences in their setting out. Such writings as the *De Arithmetica* of Anicius Boethius (AD 480–524), describing proportional systems found in nature, became sources for deriving structural dimensions. By such sources Neoplatonism maintained itself through the Dark Ages.

As Vitruvius had done for the Augustan era, the historian Procopius (AD 500–565) recorded the construction techniques of late antiquity. His *Buildings* describes the Emperor Justinian's construction campaigns in terms of their propaganda value,[26] monuments being thrown up rapidly with means developed for speed and economy. The imminent end of the world was widely believed in. Permanence on this earth was not very highly rated.

Means of communication were very efficient in the late Roman Empire. To secure his position as Caesar, Constantine was able to travel from York to Rome in less than a week and the passage of ideas was also remarkably unencumbered. The ecclesiastical standardisation of vernacular languages and the common *Koine* assisted interchanges of ideas while early scholasticism grafted an encyclopaedic cosmology onto the disparate elements of classical learning. Building design took its place in this monolith. Latin's subsequent ascendancy as the universal language of the West promoted the capabilities of a particular type of specialist and technical expression, and while on the one hand Greek subtleties of thought were lost in translation, on the other vernaculars mixed themselves to initiate new models of thought.

## Monasticism and aesthetic rules

Monasticism, Christian or otherwise, flourished, spreading order and producing establishments based on completeness and self-reliance, as if prepared for a coming storm. Practical building forms such as refectories, hospitals, granges and stores were standardised, and plans became codified as part of the general discipline.[27] Many of the great houses demonstrated in their precise layout an immutable attitude to the relationship between man and his world, while classical sensibilities persisted in the proportions and construction methods of the monasteries as well as in the contents of their libraries.

2.6
**Visigothic church of St John the Baptist, Baños de Cerrato, Palencia, AD 661**
Arriving in Spain after a long migration via northern Africa, the Germanic people, the Visigoths, took over the Roman province. Neither completely suppressing nor assimilating their hosts' culture, they seem to have selected from the technologies available to develop an austere style of simple construction techniques subtly deployed.

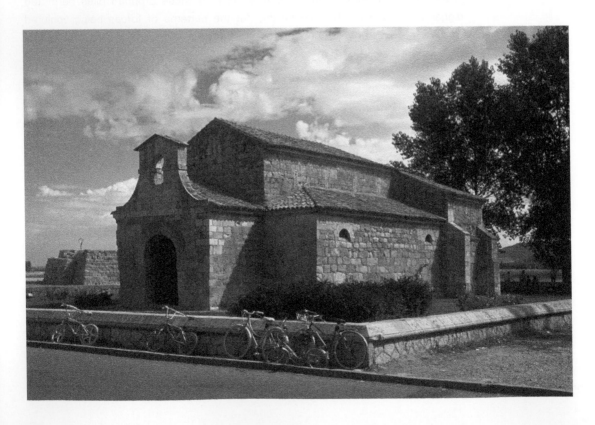

## Barbarian under-exploitation

This chapter on the embedment of classical ideals in the western consciousness concludes with an example of partial acceptance. The Visigoths wandered the empire before founding a kingdom in Spain, but their architecture is peculiarly austere: not necessarily crude but pervaded by an under-exploitation of the technology available to them from their conquests (**figure 2.6**).[28] They wanted the motifs but not the refinement that the Roman masons subject to them could produce.

Chapter 3

# Byzantium and the European Dark Ages (476–1000)

Between the ancient empires, marginal lands, deserts and mountains preserved much older ways of living. Transhumance peoples, nomads and traders lived under canvas or felt, and their tent structures perpetuated an attitude towards unmediated nature different from that of city dwellers and farmers. Maintaining flocks and moving them seasonally across altitude or latitude produced a surplus of skins and sinews that were combined with materials from the land to make shelter, and the imperative of survival set a precise balance point of lightness/portability versus robustness: no superfluities or redundancies. These forms of construction have no relationship with urbanism. The construction phase repeats over and over, the limit point of prefabrication focused around rapid and safe assembly and dismantling; almost continuous regimes of maintenance make material longevity irrelevant. Low levels of social organisation, often the family unit only, restrict component size and emphasise ways of lifting in the element design.[1]

The origins of transient construction are remote. Out of the Palaeolithic times, 12,000 years ago, came the first social groupings that pooled resources. Their shelters reflected more than the vagaries of environments. Different forms of shelter condensed around the custom and identity of the various tribes. As well as bringing practical advantage, the steady nuancing of detailing over so many repetitions produced a sign system. Nomadic buildings might incorporate some of the products of the new urbanism and its industry, such as metal fastenings and cordage, but at the same time exhibited strong resistances to assimilation. Vestigial details abound.

## Technological excess in lightweight construction

Technological excessiveness, over-refinement of design, makes an early appearance in the social differentiations between tent forms. Some structures were made needlessly complicated and protracted in assembly specifically to indicate an overabundance of human resources.

*3.0 (opposite)*
**Nomad Tent, the Silk Road, today**
A structural system predicated on transportability and resistance to wind loading. Tension elements, membranes and ties, ensure minimal weight, and overall shape determines stability. The flexible components are easily packed, repaired and replaced. The organic materials available to transhumance peoples, wool, tendons and bone, suit this construction.

3.1
**Japanese Pavilion, Kyoto, AD 750**
Log size determines the way timber structures can be built up.
The joints in this Japanese pavilion are a stylised refinement of
de-mountable pegged joints and are capable of sustaining load
reversals from inertial forces caused by earthquakes.

Many of the power structures of the early empires involved peripatetic administrations touring around new and old conquests, and whole families of structural details developed from this practice. In Japan the fighting among tribal warlords was not stabilised until the ascendancy of Yamato Japan around AD 400,[2] and the chieftains travelled with panellised paper and wood buildings, prototypes for the demountable pavilions of the Japanese court. The subjugation of the country permitted an influx of ideas and expertise. In the sixth century a small group of Korean Buddhists arrived in the kingdom of Paekche bringing along their expertise in Chinese joinery,[3] and as their beliefs were spreading so too the religious refinement of their work was adopted to become the official architecture of the Fujiwara clan. The primitive pegged junctions of earlier structures were replaced with exquisite interlocking dovetails and tenons, and Japanese carpentry's classical era had begun (**figure 3.1**).[4] The new orthodoxy of construction was embodied in the shrine site of Nara.

## Engineering for defence

Crossovers from military technology influenced the evolution of the islands' distinctive construction style. Relentless warring had promoted particular improvements in structural form. Castles comprised robust timber frames set on massive plinths of polygonal masonry, fireproofed with gypsum render, and the perfecting of archery caused the perfecting of tightly sealing shutters and close-boarded screens in the upper works of castles.

Japan is raked by earthquakes – regular tremors as well as major shocks – and structures must be especially refined and capable of repair or rebuilding, as lighter structures are subject to lower inertia forces when shaken. The early timber frame joints became unusually complicated to resist the tensions and load reversals of uplifts. Tiered forms have complicated resonances unresponsive to passing shock waves, and the bracketry of traditional Japanese roof framing contributes exceptional levels of damping to dissipate the energy of ground movements. The traditional Japanese pagoda structure – strong central core surrounded by stages of framing and topped off with early warning bells – is an almost perfect arrangement to resist vibration, and many examples have survived down the centuries.[5]

## Early European timber construction

The timber constructions of northern Europe and Scandinavia later in the first millennium had developed very differently from those of Japan. Communal buildings had long since reached a scale needing real structural frames of tied logs, set up as trestles holding up pitched rafters. In the western Roman Empire, as centralised control waned, the variation in timber types caused by climate and geology reasserted itself in the variety of local vernacular building. Paradoxically, circumstances of social deterioration allowed a more general replacement with iron of the stone and bronze tools whose use had been imposed on the underclasses. Materials hard enough to take a real edge became common and were no longer limited to weapons, the preserve of the aristocracy, meaning that the harder timbers, oak and beech, could now be worked, albeit still green. Such trees yielded large and heavy baulks. Roman knowledge of handling systems, cranes and levers, was not lost but transmitted through the construction of increasingly large structures and roofs built up of simple trestles or triangulations. Joints were hardwood-pegged, kiln-dried, driven and then watered to expand into rigid fixings. The evolution and distribution of these forms can be tracked across Europe, tracing one flow of medieval ideas as ever more refined joints filtered down to domestic construction.[6]

## Hagia Sophia, Constantinople

Trees size themselves, the best to survive, and early forestry could improve log lengths very little.[7] As the scale of buildings increased beyond single-beam spans, the times could only make up larger timber structures from elementary sub-assemblies and hierarchies, trestles and kingposts. Masonry construction, however, governed by proportions only, could be scaled up and up, seemingly indefinitely, and by the middle of the first millennium the permanence and consistency of stone and brick were being exploited in buildings of completely unprecedented scale.

There is little sign of a step change in the design approaches taken towards masonry building. This is the age when the very largest structures were erected in a way completely 'other' to our own understanding. Structures of modern scale were designed by minds of ancient outlook, whose coherence of thought would seem completely alien to us. The paradox to be explored is how these ancient engineering minds, fundamentally different from ours, achieved safety margins that we would still be loathe to reduce. Their ideas lose definition under the harsh light of modern engineering insights.

The church of the Roman emperor Justinian (AD 527–565), the Hagia Sophia in Constantinople, now a mosque, was commenced in 533, as a response to escalating civil unrest.[8] Its construction was intended to consolidate a religious orthodoxy, to embody a liturgy and cosmology that would be unchanging, and a huge effort was made to create a building that would be a true microcosm, a representation of the universe (figure **3.2**). Christianity's ascendancy in the Eastern Empire had great political significance but was marred by deep inconsistencies across its theology. The spatial qualities of the new church involve ambiguities of hierarchy and boundary, with the structure embodying a balance of forces concealed by the continuity of surfaces and the permanence of the underlying masonry underscoring the strength of the state.

Many centuries later, following Ataturk's (1881–1938) secularisation of the Turkish state in 1934, the building became a site for intense investigation, partly in aid of restoration and partly as a pleasant destination upon which to theorise.

3.2a and b
**Hagia Sophia, Constantinople, AD 532**
The delicate bubble of masonry suspended over the city depends on a whole series of structural patterns. Finely dispersed ribs around the springing of the dome allow for a band of windows. Larger buttresses below frame infill surfaces pierced to control light levels within.

The first roof collapsed in an earthquake. The replacement is carefully pitched to achieve longevity and lightness.

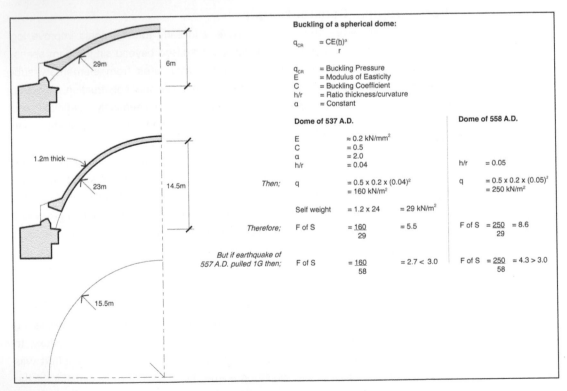

**Buckling of a spherical dome:**

$$q_{CR} = \frac{CE(h)^a}{r}$$

$q_{CR}$ = Buckling Pressure
$E$ = Modulus of Easticity
$C$ = Buckling Coefficient
$h/r$ = Ratio thickness/curvature
$a$ = Constant

**Dome of 537 A.D.**

| | |
|---|---|
| $E$ | $\approx 0.2$ kN/mm$^2$ |
| $C$ | $= 0.5$ |
| $a$ | $= 2.0$ |
| $h/r$ | $= 0.04$ |

Then; $q$ = $0.5 \times 0.2 \times (0.04)^2$
= 160 kN/m$^2$

Self weight = $1.2 \times 24$ = 29 kN/m$^2$

Therefore; F of S = $\frac{160}{29}$ = 5.5

*But if earthquake of 557 A.D. pulled 1G then;* F of S = $\frac{160}{58}$ = 2.7 < 3.0

**Dome of 558 A.D.**

$h/r$ = 0.05

$q$ = $0.5 \times 0.2 \times (0.05)^2$
= 250 kN/m$^2$

F of S = $\frac{250}{29}$ = 8.6

F of S = $\frac{250}{58}$ = 4.3 > 3.0

Diagram labels: 29m, 6m, 1.2m thick, 23m, 14.5m, 15.5m

The historian of structural engineering Rowland Mainstone and others have documented the existing remains, carefully back-analysing the movement of the structure and remedial measures that have been taken over the years.[9] A story of construction, observation and intervention has been unfolded in which we see an example of the way masonry structures enter a dialogue with their designers and reveal how they are working. This is the significant aspect of masonry construction used by the ancients. The engineers closely observed their material's behaviour: its ductility was such that it deformed and shaped itself visibly into load paths, which could be detected, mimicked or enhanced.

## Greek mechanikoi

Justinian chose two designers to create his church: Anthemius of Thrales (474–534 BC) and Isidorus of Miletus (fl. 520 BC). Both were of Greek colonial origin, a background renowned for construction know-how. In *koine*, the everyday Greek language of the New Testament, the two men were styled *mechanikoi*. The word does not just mean engineer, but implies a mastery of both practical craft and theory in a wide range of topics. As conceptual designers, they were supported by a small army of *architektoi*, practical builders, leading the construction teams and day-to-day detailing.

　　　　Both men taught and wrote on geometry. Anthemius was more the mathematician, a writer on conic projections whose skills at setting out were offset by his ignorance of building craft – a combination suited to lateral abstractions. Isidorus was an experienced architect and builder, capable of reining in ideas and delivering buildings. He produced a commentary on a work now lost, *On Vaulting* by Hero of Alexandria (AD 10–70), a book which would have discussed setting-out problems, estimates of quantities and volumes related to Roman vaulting. Concepts of statics appeared to be limited to Archimedes' lever and a notion of the centre of gravity of an object. Archimedes had downplayed the practical applications of his discoveries, and it needed hard-nosed builders to overcome this resistance. That stability depends on weight distribution, thrust balancing thrust, is an understanding demonstrated by their use of a range of weight-reduction devices, but they worked uncertainly. Supporting piers, originally pierced, were filled in to improve lateral stiffness. It was said that special lightweight bricks were fired on the Island of Rhodes for the main arches, but subsequently they have proved to be of normal density. The structure did incorporate many different materials, however, for the entrepôt of empire could exploit cheap and remote prefabrication and acquire rich materials. Components of ancient ruins were subsumed, such as columns from the temple of Ephesus. A construction period of only five years was claimed.

## Transitions expressed in built form

The designer's desire to sublimate conflicting spatial ambitions drives the setting-out of the church. To describe the myriad patterning as a transitional form between basilicas and centralised plans is to shoehorn it too hard into a development trajectory. The interlocking spherical forms were first and foremost a geometrical resolution, combining aisled spaces and centralised form; the structural benefits were secondary. A conventional classical construction of piers strengthened with tiered arches was recast into a 'cascade' of elements taking spreading forces from domes and half-domes outwards and downwards onto widespread foundations. These patterns of force are perceived at an instinctual level. The massive piers carrying and confining the main dome are disguised between the receding spaces and brightly lit tympana, so that the whole interior almost floats on the mote-filled air.

## Building on precedent

The originality and sheer creative tension evident in the design of the great church scatters its antecedents, but the layout appears to have been prefigured in the nearby, much smaller, church of St Sergius and Bacchus.[10] Differences alongside similarities highlight the designer's preoccupations. Tolerances on the setting-out of the larger building, to which the structure would have been very sensitive, were tightened up considerably. Several lesser Byzantine churches incorporate innovative devices that were not scaled up in the bigger building. Tension belts of stone cramped together with metal ties girdled many domes to control spreading without massive buttressing. Started six years before Hagia Sophia, the wondrous church of San Vitale, Ravenna, took centralised planning to an even higher pitch of spatial complexity.[11] The beautiful cupola is ruched to stiffen the eggshell-thin surface and incorporates interlocking earthenware and pumice blocks to lighten the vault concrete.[12]

## The stutters of technical change

These pages have emphasised the sporadic and provisional development of early structural understanding. New challenges provided the impetus for change. Genuinely unprecedented problems presented themselves only occasionally through combinations of particular circumstances.[13] The will towards perfection and completion only comes much later. Today's structural understanding retains vestiges of this episodic development. The Eastern Empire mixed up a mainstream of peoples and ideas as well as supporting niches where marginal developments could flower. Far more slender than frontier watch towers or castle bastions, early Christian campaniles – towers for bells or semantrons (wood sounders) – were attenuated to the limit where for the first time lateral stability, and overturning, became the dominant structural concern. Sited on the weak alluvial soils of cities established on river banks or coastlines, these brick and stone shafts shifted and settled, and soil-structure interactions revealed themselves as quantifiable phenomena. Pattern belied the unfathomable act of God; the propensity of the ground to buckle and of a shaft to lean could be meaningfully studied as a geometric problem. The spread of Islam was to bring even more slender minarets across the Near and Middle East.

## Arab thought

The jump cuts in thought caused by the Arab conquests of the seventh and eighth centuries AD are fundamental to the West's current outlook, and the complexity of changes makes it difficult to generalise. This has been cast as a process whereby classical philosophy was husbanded and developed by Arab scholars to be reintroduced to the West as an escape from the thickets of scholasticism, metaphysical systems veneered onto Christian belief.[14] Suffice to see here that along with this ebb and flow of ideas came re-interpretations of engineering know-how, Archimedes' screw, levers and geometrical forms,[15] as well as mental constructs such as algebra (al-jabr – 'completion'), the mathematics of abstraction.[16]

Despite Christianity's ascendancy in the West, the pagan outlook enjoyed more continuity than was once acknowledged; animism still resided in the craftsman's feeling for material and classical ways persisted. In the drawn-out confrontation between East and West following the Arab conquests of the seventh and eighth centuries, rather than a recovery of lost material there seems instead to have been a western reassessment of familiar understandings in the light of alien but extremely successful interpretations of the same canon by the Muslim world.[17] Neoplatonism's practical outcome appears to have included a re-centring of the universe on humanity and its potential to achieve, with technology, was demoted from revelation to artefact, just a tool to be used. Construction methods could be manipulated and inflected, proportions varied and vernaculars mixed, to generate hybrid structures. Underlying forms were more readily revealed as differences were no longer suppressed.

## Duality in western consciousness

The two routes by which classical thought has passed to us leave it very open to interpretation,[18] and this porosity also permeates Christianity to its very core. The way ideas are absorbed continues to shape our development. We still project a concealed structure behind the surface of things. The Arabic interpretations of western philosophy simultaneously gave it a provisional quality, which has allowed us to add to and alter a contingent edifice ever since. Abstraction and algebra took on the dominance they retain today.

## Mediterranean marginalia

Drawn along by the weakness of its neighbours, the Islamic conquest flowed out of the Arab peninsula and around the shores of the Mediterranean. An interface of cultures stretching from the Middle East, across Sicily, Rhodes, north Africa and into Spain produced a margin on which many ideas significant to structural engineering condensed. The ruins of massive foundation plinths and aqueducts record Roman construction expertise in Cordoba, Spain. The city's Mosque, commenced in 784, incorporates more than five hundred columns, many taken from Roman and Visigothic monuments, imperial ruins.[19] The Syrian architect Sidi ben Ayyub compensated for varying column heights by combining the Roman structural system of tiered arches with an Arabic invention of extended impost blocks, arch bearings. The ancient porticos are resurrected in a completely new cosmology, crystalline and infinite. The logic of classical Islam is perfectly reflected in these early buildings. Local forms were appropriated, then recast to reflect the rationalising power of the incoming religion.

## Ancient Chinese technology

At the same time that Rome was founded, throughout its empire and as it foundered, another great civilisation shaped itself. Over more than a millennium after the second unification of China by the Qin in 221 BC, the Great Wall was assembled from piecemeal fortifications and extended into a civic structure unifying a country. Physical defences were backed up by policies of trade and appeasement towards aggressive neighbours, and the peace and shelter provided by the wall permitted civil technologies to develop. The hordes beyond the western and northern borders were to be bought off, and this required a vast central administration to be run on a developed taxation system: paper and writing for accounting, an infrastructure for distribution. A network of canals supplementing the great waterways of the eastern and central areas provided for the transportation of bulk goods while roads extended communications and took armies westwards into the mountains and deserts that formed the empire's borders. Once the political necessity of paying tribute to the barbarian invaders had been accepted, goods flowed along these routes – up to a third of the empire's gross product.[20]

The scale of this logistical exercise paid for steady technical improvement. Appeasement goods set out from their lowland sources on barges drawn upstream. Vine and hemp tow-ropes were replaced with metal chains made of wrought iron links, a by-product of developments in metallurgy necessary to refine swords. Strong tensile, elastic and ductile metal replaced brittle cast iron.[21]

Ancient China's trade link westwards along the Silk Road skirted around the Gobi Desert, the dispersed trails winding through the foothills north of the Himalayas, or south of Mongolia. Progressing into the highlands, cargoes were baled and carried on pack animals or human shoulders. This dispersion of loads could then be led across primitive rope bridges slung over precipices and ravines gouged and boulder-choked by spring floods. Light bamboo decks and hemp cables of consistent strength secured to firm rock abutments combined to make suspension bridges of great spans. On well-trafficked routes the reliability and longevity of forged metal encouraged the replacement of rope catenaries with superannuated towing chains.[22] Fibre bridges, instinctively set near the ideal sag ratio of 1 in 10 by their builders, rising and falling with humidity changes, could make spans up to around seventy metres, but metal suspenders, much stronger and stiffened by their own self-weight, made safe crossings of more than 200 metres, nearing the practical limit of a modern stressed ribbon footbridge (**figure 3.3**). The inherent flexibility of suspension bridges also led to their use in multiples across wide deltas and marshlands, easily accommodating the large foundation settlements of the weak alluvial soils.

## Bridge forms developed in parallel

The Chinese developed several bridge types in parallel. The empire's variety of geography, materials and traditions generated diversity, and its bureaucracy and centralisation promoted standardisation. Forms with different origins were improved to a consistency to provide a lexicon applicable to specific problems. Canal-side

3.3 (*left*) and 3.4
(*right*)
**Luding chain
bridge, Dadu river,
Western Sichuan,
1701, and
Ghengyang timber
corbel bridge,
Linxi
river,Guangxi,
1916**
The Han Dynasty
developed several
bridge forms,
maintaining a
flexible engineering
approach to
different conditions.
Lightweight chain
bridges suited the
wide river deltas as
well as the deep
mountain gorges.

villages needed short, steep-sided bridges, so stone arches marked village centres.[23] Itinerant masons used their repetitive trade to refine a standard solution. The overall height of the crossing was minimised by pinching the depth of structure at the arches' crown. Low-profile elastic rings of masonry with carefully distributed spandrel loadings exploited the ready-made abutments of the canal bunds. The thin sections were able to cope with differential settlements that would destabilise thicker less flexible arches, and on wider spans over rivers pierced spandrels were introduced to relieve flood-water pressures and reduce top weight.

The timber alternative to these stone canal bridges, the 'rainbow bridge', has, in the absence of archaeological record, become a site for imaginative reconstruction. Images on antique porcelain omit sufficient detail to admit speculation. How such bridges were assembled is for conjecture.[24] What is known is that low segmental arches were formed by interlocking sets of manhandleable timber segments. Here was an embodiment of that Chinese ideal of the whole being more than the sum of the parts. The same principle informed a more primitive precursor: the corbel bridge (**figure 3.4**). These ancient structures have survived, not only as a result of robustness but also because of their practicality of maintenance and repair.[25] Flood damage and decay is dealt with by the replacement of individual components without dismantling the entire structure. Like the proverbial grandfather's axe, the structure persists longer than the characteristics of its individual parts would otherwise allow.

## Chinese boatbuilding: bulkheads

Along the Chinese seaboard contributions to structure were made through a different set of technologies as trade upon the wide deltas of the Pearl, Yellow and Yangtze rivers spawned a whole range of coastal and sea-going vessels. Different tribes produced different craft, with distinctive characteristics reflecting purpose and material availability.[26] Successful trading prompted partnering and joint ownership of boats, and in order to separate the goods of each party, hold dividers began to be inserted. These bulkheads simultaneously strengthened the hull and compartmented it, offering the possibility of staying afloat even after inundation by storm or below-waterline damage. Most modern monocoque structures keep to this simple pattern, and it subsequently became the preferred approach to ensuring robustness in all kinds of structure. Junk sails, low-technology assemblies of hemp, jute and bamboo, were nuanced to the limit of their potential.[27] They are responsive structures of great sophistication capable of control by a single-family crew.

The lightness of boat construction equates to a cultural attitude. Around the world from China, the environment of Scandinavia and the Baltic Sea focused boat-building along a single trajectory. The Northmen made beautiful ships, lithe but incredibly sturdy, from the slow-growing, rock-hard, straight pine of the winter forests (**figure 3.5**). Their hull structures met conflicting demands. Clinker construction compressed and sealed the planking joints as well as distributing the high bending stresses induced by the long wavelengths encountered in the North Atlantic. Boats could be practically and rapidly built and repaired on any piece of flat ground. By this means of structuring, they were light and strong enough to be landed at night, which limited water-logging, and could be taken overland.

Communication eastwards was hampered by forests, wide expanses of glacial till, innumerable lakes and the succession of rivers flowing northwards. The wide shallow hulls could be dragged across the open pine floors from river to river and lake to lake. Following a lucrative slave trade, Nordic explorers and raiders used

their boat-building skills to pass down through Russia as far as the Arab world and into the Mediterranean basin. Westwards, the good sea-holding characteristics of the hull lines achievable by clenched planking, fitted with more generous freeboards, allowed their builders to venture on to the North Sea and Atlantic. Traders, raiders and settlers, they reached westward to America, and southwards to Britain, Ireland and France.

The archaeology and literature of the northern lands includes an idealism, almost romanticism, in attitude towards this technology. The Osberg ship is a burial offering, recovered in pristine condition, a full-scale vessel but of idealised form, practical yet of such low freeboard as to be unfit for the open sea; exquisite in pro-portion and construction, the cart included among the grave goods on deck is useless. The components of the ship are extraordinarily attenuated, almost transposed into a dream world, and the formal terminations of the keel, stem and stern, are sublime.[28]

Alongside the sophistication of the Scandinavian shipwright's timber technology, a cruder building vernacular was maintained. The early stave churches that appeared with the coming of Christianity are highly decorated but devoid of technical excess. The liturgy called for a congregational hall and subsidiary spaces surrounding a sanctum, and pagan motifs and devices were mixed with these requirements to produce structures with a consistent complexity extending across form, spatial ordering, jointing and detailing.[29]

## Atrophied technologies

The pre-eminence of boat-building and water transport seems to have atrophied other developments. Infrastructure was almost unknown. The archipelagos and lakes of the glaciated Baltic landscapes had favoured water-borne trade; eventually the eskers and terminal moraines became routes overland. Northern chieftains advertised their power by establishing, maintaining and controlling causeways and bridges. Their ships were decked, but only slowly did suspended floors begin to appear in halls and aristocratic houses. The technology for the long trestle causeways approaching the major population centres, ideal tax points, had to be imported from the neighbouring Slavs.[30]

The expertise gained in establishing transport links transferred to the establishment of extended defensive works, walls combining defence with major communication routes. These heavy engineering works required improved levels of organisation. A series of quite extraordinary round forts appeared in the south of Sweden and Denmark.[31] Their layouts mix patterns of military organisation with Roman units of measure and geometric ideas possibly from as far afield as the Arab south in an experiment that was extraordinary and short-lived.

There is a time for diversification and a time for consolidation in materials development. Metal production and smithing in ancient Scandinavia was dispersed and disorganised, broadening its range of development. This dilution of skill among farmers created an imperfect practice of the craft through which improvements could appear. These alternatives could be rapidly disseminated by itinerant smiths and through the small products – brooches, strap hinges, hooks – that raised the money system just above barter. Bracelets of silver and gold, of standard weight and

3.5 (*opposite*)
**Osberg Boat, Oslo, AD 800**
The glacial lake-lands, archipelagos and forests of Scandinavia perfected wooden ship-building. Shaped to meet the North Atlantic, the hulls relied on clinker construction, lapped side planks clenched with nails. Joints were watertight, transferring shear forces without sub-framing. The lightweight shells proved simple to build, robust enough for beaching and portage. Part of a woman's grave goods, this vessel was old at the time of its interment. It is an idealised artefact, the freeboard and lines too low for the open sea. Functional elements and structural components have been refined almost beyond the point of practicality.

type, were an established repository of wealth, and base metal exchanges were a day-to-day support in many of the antique economic systems.[32] The widening number of sources for ore meant that different amalgams and alloys had to be understood and manipulated.

Resistance to Viking raiding gradually coalesced. The countries and trading states of the Atlantic seaboard put to sea in convoy, and confrontations between fleets began to occur. Strength became more important than speed. High-sided vessels, particularly the 'cogs' of the Hanseatic league, tended to prevail in these encounters and began to evolve into a new kind of fighting ship based on height, size and robustness.[33] Unlike longboats, their main deck was an integral part of their structure, contributing directly to hull stiffness. These were the first true monocoque structures, safe on the open ocean. Sail-handling had to improve correspondingly to keep these ships practical and manoeuvrable.

## Landscape engineering

The Vikings had little use for heavy engineering and large structures. The examples cited so far in this chapter – buildings, bridges, boats – all have structures that can claim to have been shaped by functional requirements mediated by other influences. Are there other sources from which relationships of structure to building have been created in a different way?

The Vikings had a very functional technology for their seafaring, one capable of carrying their cultural identity. Their borrowings for building and infrastructure were left as forms uninflected by that spirit.

Elsewhere, conditions initiated a strand of engineering whose origins are agricultural. The structures of 'landscape engineering' shape and are shaped by the environment, and in Central America the early civilisations made engineering reclamations in the fertile lake areas and plains. Field patterns became urban outlines. Powerful bureaucracies provided the organisation to build artificial hills supporting religious ritual. Despite having very different functions and build programmes lasting generations rather than single lifetimes, these mounds have internal structures, layered like onions, entirely confluent with Egyptian examples built 3000 years earlier. Structure close-coupled to form.

In south-east Asia the vast temple complexes of the Khmers encoded a process whereby agricultural patterns were first adapted to fortification and then imprinted and transformed by a succession of religions.[34] The great site at Angkor, continuously occupied from the ninth to fifteenth centuries AD, shows little technical development as the styles change from Classical to Baroque phases and from one religion to another (**figure 3.6**). The Prasat towers are made without voussoirs and instead are corbelled in an ancient way to produce false vaults. Towers built over successive generations never vary from proportions deemed to be safe. Structure vestigial to form.

In the most extreme examples of landscape manipulation, structure evaporates and tectonics have no influence on scale or proportion. The stone massifs

**3.6**
**Angkor Wat, Cambodia, 1130**
The building campaigns of the temple complex served a number of religions. Stylistic developments, classical through to baroque, have been traced by commentators, but a continuity of technique is maintained across each 'jump cut' of culture. This is due in part to a caste system isolating workers, in part to the momentum of building technology itself.

of the Himalayan fringes were adapted to contain complete, vast buildings carved straight from the living rock.[35] Temples and pavilions shaped from the outcrops or deep halls and niches hollowed into the solid cliffs took on any form that could be imagined, and references to conventional building disappeared. Form without structure.

# Chapter 4

# Light (1000–1600)

## Christianity in Northern Europe

By the end of the first millennium in northern Europe, all thought was controlled by faith. The Christian religion was supercharged with potentials for change. Despite conflict and plague, the High Middle Ages saw periods of prosperity and economic expansion, sufficient to support strong efforts to put the social order on sound theological footings. While church and state competed for power, the internal inconsistencies of the Christian faith, seemingly irreconcilable, kept ideas fermenting. Scholasticism, an attempt to use precedent and church authority in a rational way, was proving unequal to the classical philosophy flowing back from the Muslim expansion while at the same time increasingly strong economic and technical bases were improving but being under-utilised. The eventual partial resolution of these problems, philosophical, religious and economic, resulted in the modern nation states. The West's reliance on technology was established. The 'modern' conception of man's place in the universe, an independent agency with free will, was created specifically as an alternative to the philosophical problems of this period.

## Saint Augustine

Augustine of Hippo (AD 354–430) lived a long and eventful life.[1] Born in Numidia to a pagan father and Christian mother, he dabbled in several of the cults then circulating around the periphery of the Roman Empire and set a pattern in claiming misspent youth before conversion to Christianity. Eventually he became the Bishop of Hippo, stationing himself near his birthplace on the north African coast, away from the establishment centre. He then proceeded with a prolific writing programme centred around the intention 'to graft Platonism onto the incomplete tree of Christian theology', and his systematic approach and clarity of thought led to his rediscovery by late medieval thinkers. Perhaps beside a study window overlooking the Mediterranean, he came to equate the holy spirit with light, an idea abstracted from various eastern religions – Zoroastrianism, the *Sol Invictus* of Aurelian. He preached the notion of light filling the soul, recasting St Paul's conversion and other stories to emphasise illumination.

4.0 *(opposite)*
**Ely Cathedral lantern, 1328–1342**
Daylight meetings, with no censers as fire hazard, allowed the cathedral chapters to be housed in timber-roofed compartments. As the problems of polygonal jointing were solved, so large frames supporting glazed lanterns could be developed. At Ely carpenters extrapolated their experience to top-light the church crossing with 40 tonnes of oak framing.

In northern Europe five hundred years later, this theology manifested itself in a new form of building, meteoric in its development. Why this late gestation? The idea and associated building technology had been continuously available in the interim but their synthesis had to be a solution arising out of a 'problematic' set of conditions.

## The manipulation of light

Bernard of Clairvaux (1090–1153) ('Clairvaux' = 'the valley of light'), a literary genius, called for a reform of sculpture and architecture, away from the popular grotesques adorning church capitals and back towards a more austere classicism.[2] All art was to be completely subsumed to the requirements of theology. Translated into built form, these abstractions had profound effect. The eastern church had adopted the basilica, with its long halls and their narrow slot clerestories sheltering the congregations from glare, while further north such construction provided the shadowy spaces more suited to mysticism. Building technology had not been challenged significantly since the social programmes of the late empire, but now interiors were to be flooded with light, God's illuminating reason. Window walls grew in scale and delicacy; the rose windows of church transepts and West ends changed from coarse radial patterns into filigree, the ever finer tracery requiring ever finer stone, tools and techniques.

## Gothic vaulting

Strict church rules stretched the observances beyond sunset and before bleary dawn, when ranks of candles and censors raised the risk of fires. The high incidence of these acts of God together with an emphasis on the permanence of the Kingdom of Heaven led to the lining of timber roofs with light stone vaulting. These thin masonry shells, stiffened with ribs no thicker than window mullions, were very different from the massive barrel and cross vaults of the old basilicas. These ceilings exerted minimal forces, horizontal and vertical, on their supporting walls so that clerestories could be forced ever higher. If Egyptian and Greek temples are homologues in stone of organic prototypes then the Gothic nave recalls the primitive grove but without an earthly precedent. Classicism is truly transformed in this masonry, re-infused rather than remnant.

The development of High Gothic ideas, in architecture and sculpture, can be followed on one site through the successive campaigns to construct Chartres cathedral.[3] Pagan motifs withdrew behind crisper lines and structure becomes steadily more attenuated while new structural devices, ribs and bosses, appear as disturbances within a steady improvement of Vitruvian building expertise.

The way in which Gothic vaulting developed doesn't quite fit a linear model of technical determinism.[4] The appearance of the pointed arch, from 'somewhere east', is the breakthrough application making possible rectangular bays in cross-vaulted ground plans. Decorative ribbing, again originating in the Muslim world, starts out as a structural framework turning into decorative motif then returning as an interlocking, elastic skeleton within which a stretched membrane of stone could be inserted. The combination of line and surface, and the embellishment of nodes with a coordinated narrative, transubstantiates the stone and produces a kind of 'deconstructed' space.

Larger timber frames were usually laid out on the ground then lifted up in rows. This process allowed relatively small social groups to make large buildings and gave agricultural structures their characteristic rectilinearity.[5] It established the visualisation of structure in single planes long before drawn representation made the two dimensions of the paper surface dominant. Masons traced out their cutting problems full size on flat plaster-washed floors. The problems of roofing chapter houses and transepts were not so easily simplified and centralised plans were covered in a variety of increasingly ingenious ways.[6] The mason developed the art of stereotomy, planning joint surfaces, assuming the use of 'free stone', masonry of uniform behaviour in all directions. Timber is strongest along its grain, and in the three-dimensional joints of wooden space frames the carpenter met the added complexity of manipulating axes and perpendicular planes to transfer forces through adequate areas of end grain.

## Beauvais Cathedral

The cathedral-building campaigns in northern Europe between 1050 and 1350 make a crisply defined, saturated problem for anthropological explanation.[7] Religious unease was compensated for by the instigation of hugely profligate building programmes, representations of huge instabilities in the group consciousness. For the first time construction technologies were pushed to the limit of their potential for idealistic instead of military reasons, and concerted effort by the entire social body committed its surplus into extreme and idealist technical endeavour. The culmination of these initiatives was the choir at Beauvais, begun in 1247.[8] In a small wool town north-east of Paris the church, never completed, stands with a clear internal height of forty-eight metres; everything is subsumed to a drive upwards into space. The walls are thin screens lining a layered system of flying buttresses and piers (**figure 4.1**). A modern structural analysis by the engineer and theoretician Jacques Heyman (1925–) reveals the stresses in the stone to be low.[9] However, the overall stability of the masonry skeleton and its interaction with the soil formation beneath is right on the practical limit that we would accept today. Projecting rather modern forensics onto one of the famous collapses of the Beauvais vaults (the existing structure is the third restart), Heyman proposes a complicated failure process escalating from a small 'trigger' incident. Implicit in his explanation is a rather touching confidence in medieval masons not to have simply taken the overall system too far into areas of unforeseen sensitivity in their first attempts, and also that the seemingly less significant parts of a medieval cathedral are structurally critical. There would be no superfluities. This might be an anachronistic application of 'Ockham's razor', an idea from around this period which has been extremely important for the development of engineering ideas and which still holds sway.[10] A widely held notion of divinity in simplicity was recorded by William of Ockham (1288–1348). As a corollary, he suggested that the simplest answer to a

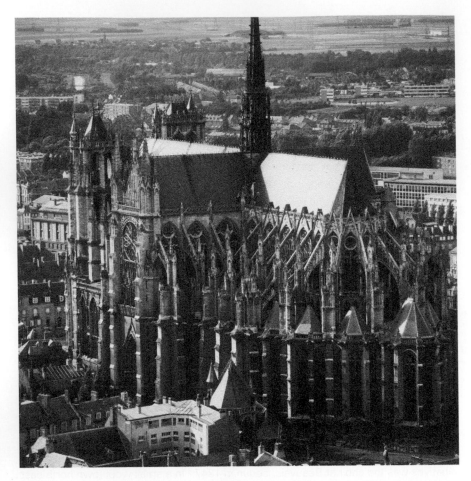

4.1 (*left*)
**Beauvais Choir,
1225–1272**
The causes of the
collapse of an area
of vaulting in 1284,
twelve years after
its construction,
have been
extensively
debated. The statics
are correct but the
buttresses have
been made so
slender that
'secondary' effects
have become
important. Iron rods
connecting the
frames were
thought to be
superfluous but
when removed
allowed wind-
induced vibrations
to damage the
masonry.

problem will always be the correct one; nature does not complicate. This idea offers engineering its moral imperative to use the least of means.[11]

Beauvais Choir is one extreme, but there were other technical excesses of which the over-extension of vaulting typologies into one late form, the fan vault, is fascinating and instructive. Of unparalleled beauty, these forms are flawed and not 'pure' structures at all. Inverted trumpets of stone rise up from slender springing points; ribbing disappears and surface becomes incised with tracery. The shells of structure intersect on arbitrary divisions and could be infinitely extended. In earlier examples these inconsistencies of form were suppressed or ignored behind the decorative emphasis, and the introduction of central spandrels mitigated some of the junction problems. Secondary, suspended conoids insinuated into these areas of flat infill produced fan-vaulting's masterpiece, the profusion of Henry VII's chapel at Westminster Abbey (**figure 4.2**).

Monasticism was a fundamental component of medieval religion. The orders crossed borders and fostered internal colonisation with establishments in wild and marginal areas. The internationalism of Bernard's Cistercian order ensured that his ideas spread rapidly. Widespread theological debate, in a Latin modified for the purpose, spread education and a technical language with which to transmit it and, like burrs on a hide, technical ideas came along too. The ability to travel, paradoxically coupled with the difficulties of communication, meant that isolated experimentation proliferated and that successes then disseminated themselves rapidly, adopted and refined.

Bernard of Clairvaux also found time to preach the disastrous Second Crusade, another way for information to travel.[12] That conflict of world views was more complicated than modern conflict with its utter rejection of enemy values.

**4.2**
**Henry VII Lady Chapel, Westminster, 1503**
Fan vaults take form beyond the logic of structure. Inverted trumpets of stone intersect, leaving discontinuities and flat, unsupported panels. A complete arched system is needed above the soffit. The complicated internal modelling of late Gothic vaults often conceals a simpler realisation of structure above. At the same time it appears that lierne vaults, patterned with intersecting ribs, were made both as stone skeletons with masonry infils and as stone shells with attached ribs.

Between bouts of appalling violence there was much assimilation and interchange of ideas. Given the mixed fortunes of the initiatives, Crusaders inevitably returned steeped in alien ideas, precipitating changes in outlook and cross-fertilising the technologies of West and East. Small elites were moving through established systems. Castles were needed to maintain tenuous footholds with minimal resources. These were not overblown power displays but grim, completely utilitarian forms. The military orders grafted monastic rules onto temporal power. Medieval fortification reached its zenith in their strongholds. The Krac des Chevaliers by the Hospitallers (1170–1250) epitomises the combination of heavy engineering and organisational complexity in these buildings.

## Mason's rules and Euclid's Elements

What sort of practical approach to design could a medieval builder fabricate from this variety of sources? Masons developed sets of rules whose analysis offers some insight into their consciousness. The prescriptions of the monastic orders, the plan layouts of their buildings, were believed to reflect deeper patterns of creation. Similarly the geometric rules in Euclid's *Elements* and the system itself were held to be superficial revelations of the universe's real but hidden structure. For builders, geometry above all else was valued as the most ready route to permanence and stability. This emphasis promoted spatial and tectonic complexity and atrophied real structural understanding. Experiments with space, interlocking and extending, and the elaboration of elements and junctions, columns and capitals, took precedent over structural invention. For the sizing of elements proportional rules learnt by rote were usually followed: instructions from which safe built form would automatically be generated. The masons invented a 'back story' that Euclid had been a clerk of Abraham during the confinement of the Jews in Egypt,[13] and this elevated the worthiness of their craft and also equated masonry work with geometry. This understanding of 'Euclidean geometry' is different from the ancients' or our own. Neither the shadow of a hidden perfection nor a generative system, their interpretation gives mental tools and hand tools one nature in common.

The transmission back to the West of Euclid's lost *Elements* came in two phases. This process bequeathed to us an acceptance of engineering's patchwork co-existing within a universal system. A partial translation (axioms without the proofs) had long been available in Latin from the Roman philosopher Boethius (AD 480–524) and others;[14] then in the twelfth century fuller translations from the Arabic appeared. Meanwhile, other sources had filled some of the gaps. A whole host of handbooks on practical surveying dropped Euclid in among their pages, treatises by the late Roman authors Martianus Capella, Cassiodorus (AD 485–585) and Isidore of Seville (AD 560–636).[15] The famous Codex Arecierianus written in the sixth or seventh century, which preserved fragments of geometric theory from the classical writers Frontius, Hyginus, Balbus, Nipsus, Epaphroditus and Vitruvius Rufus, resurfaced in the tenth century in the monastery of Bobbio. These hand-me-down collections with gaps in between were suited to arcane and esoteric interpretation. Their openwork of knowledge made room for a practical approach to building to appear.

How much practical geometry did the masons draw from the academic treatises assembled from the rediscoveries? Isidore of Seville had defined geometry as the 'measure of the earth'. This ancient concept of geometry as manifestation of the cosmos now slid into geometry as tool with which to reproduce form. University geometry retained the old purity so the mason's understanding of the subject had already undergone at least one transformation to become the craftsman tool, encoded into a tradition system to allow safe transmittal. Euclid had proposed that all his axioms should be readily constructible in diagrams only as a control on his thought. Obviously such a stipulation was fundamental to the *Elements* used by the medieval mason. This geometry was 'constructive', generating solutions from the manipulation of compasses and ruler only. Real number properties were irrelevant. The absence of proofs in the early texts is telling. Mathematics was becoming detached from practical usage, but more importantly the two were being conceptualised differently as well.

Design information had to be transmitted, taught to others and simultaneously protected as a valuable asset. In my opinion this antipathy at the core of structural engineering is still its defining characteristic and major limitation. The complicated fences medieval masons used to protect their knowledge arrested its development and at the same time instituted a prototype form of quality control. Geometric understanding had been reworked and was to be transferred by word of mouth and physical demonstration, coded and stored accordingly.

So the corpus of medieval construction knowledge was deep-structured by the conflicting demands of concealment and dissemination. The difference between literacy in Latin and in the vernaculars is important. The 'quadrivium' was a set way of organising knowledge used by the Latin schools, strictly defined and immutable. Arithmetic, geometry, music and astronomy did not meld in the crucible of practical use as they might otherwise have done.

## Systematic knowledge

But within the straitjacket of received knowledge books searching for system appeared, system distinct from ancient authority. One early example of the old knowledge not merely being copied but rather analysed, organised and synthesised is found in the compilation called the *Geometria Gerberti*, with Gerbert (946–1003) as editor rather than sole author. What this monk from Aurillac, eventually to become Pope Sylvester II, proved capable of was formulating new problems from the received axioms, and broadening their scope in the process. In his re-working of the *Elements* he provided prototypes for the two main approaches to geometry in the Middle Ages. He systematically worked through the consequences of the rules, applying them to everyday stone-cutting problems. He re-attached mathematical proofs and their corollaries to the solutions he had found.

Despite this particular initiative, the split between theory and practice continued to widen. In Hugh of St Victor's (1078–1141) treatise *Practica Geometriae*, made a century later, pure geometry was dispatched to the realm of metaphysics and practice reduced to the mere manipulation of instruments. Hugh's writings demonstrate a profound mysticism. Paradoxically his original contribution was the idea of geometry's universal continuity, treating local and global measurements as one thing. The arbitrary subdivisions of his text seem to follow an accepted thought structure. Were these divisions a subconscious acknowledgement of the dislocations and inconsistencies of the mason's project?

What is important about these times is the variety of positions taken by commentators. Some tried to find theory for all practice; others rejected theory altogether, piling usages together. Monographs appeared focusing on specific problems in great detail then embroidering variations on the theme. Writings explored the need for accuracy. Sequences of pamphlets show quite how mutable craft tradition could be over time and interpretation.

Firstly, there were those who sought to meld theory and practice. The important characteristic is the process of grafting two sides across a gap rather than generating seamless system from a single set of principles. Dominicus Gundissalinus (fl. c. 1175), a native of Moorish Spain, mixed the ideas of Hugh of St Victor with those he translated from the philosopher al-Farabi (870–950). The Frenchman had made geometry mystical, the Arab held that the way to truth could only be through the interpretation of symbols. Now the Spaniard conjoined makers with measurers; every-thing can be measured, assimilated and recreated through geometry. Gundissalinus acknowledged the separateness of theory and practice but diluted it out of existence by vastly increasing the specific cases that flowed from each generalisation.

The Italian mathematician Leonardo Pisano (1170–1250), nicknamed Fibonacci, went touring around the Mediterranean world to meet the Arabic scholars. His book *Practica Geometria* maintained a separation between theory and practice, but the particular importance of the document lies in the clear exposition of how practical techniques can be derived and proved from theory, a connection so important to modern practice. In an anonymous treatise entitled *De Inquisicione*, the mathematician author is seen constructing solutions out of the axioms, thereby claiming an unprecedented accuracy. At the time masonic handbooks did not differentiate or relegate approximate solutions from exact ones.

## Villard de Honnecourt

Manuals rejecting theory are in relatively short supply so might not have been widespread. A confused little project, the anonymous *Pratike de Geometrie*, sets out practical surveying using geometry without making any of the cosmological connections usual back then. The book is in the Picard dialect, like its contemporary Villard de Honnecourt's sketchbooks, the most famous of the medieval treatises (**figures 4.3** and **4.4**).

**4.3**
**Villard de Honnecourt's portfolio, 1230**
Set alongside directions for more prosaic construction problems, usually given in recipe form without background or lengthy discussion of underlying forces, are a series of playful sleights. These ideas improved structural understanding, pre-adaptations for developments to come. The importance of the voussoir joint angles is emphasised in the sketch.

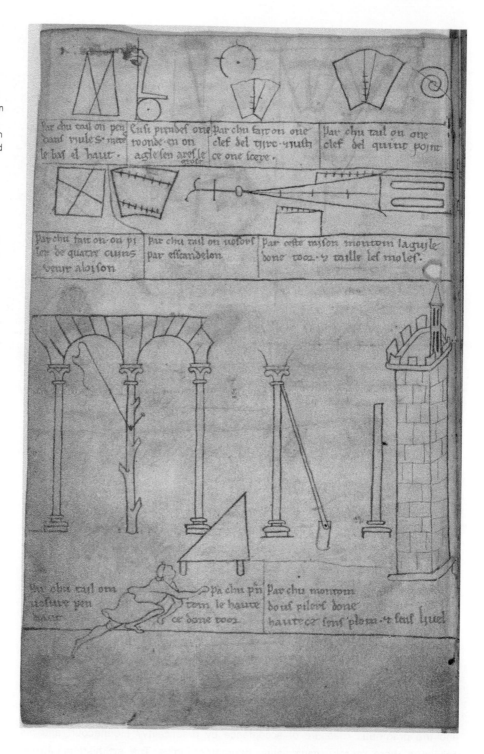

These folios are an amalgam of contributions from the named author and an unknown associate. (Researchers call him 'magister 2'.) Apart from a few plagiarisms from more academic treatises, most of the problems treated seem to stem from the direct recording of an oral tradition and the inclusion of unresolved puzzles supports this supposition. Often the narrative contracts to an aide-memoire. At one level it is what Honnecourt and Magister exclude that makes the work so informative, for they either weren't abreast of the Arabic introductions or ignored them completely for some reason. For them, geometry is nothing more than a set of tools to enable drawing and setting-out at a practical level. The techniques given for sizing elements are rules of thumb, step-by-step manipulations of instruments with little or no mathematics involved. It is unclear whether the authors actually grasped any of the principles involved and it is difficult to recover the derivation of some of the rules. Is this the starting point for our own indifference to underlying assumptions and the pragmatic measure of acceptance by practicality? Honnecourt and his colleague were merely manipulating geometric forms. The art historian Erwin Panofsky notes the limitations of his measuring systems to accurately describe natural organisms or found artefacts.[16]

4.4
**Canterbury Cathedral Great Cloister, 1400**
Structural tricks could be harnessed to architectural effect. The hanging keystone trope is here combined with filigree stonework and slender turned pillars. The attenuated structure relies on the surrounding mass for its stability. A weightless sun-screen is created, a subtle tissue along the boundary between the worlds of cloister and garden.

What the sketchbooks do most successfully is reflect a fundamental shift in design approach. Ancient buildings had been generated by encoded proportional systems that ensured a harmony of parts and a simple stability.[17] The complexities of the new ecclesiastical architecture took drawing, pre-adapted as a recording system, and changed it into a tool for the new work. Villard's sketchbooks appropriate the form of the classic building manual but shift the emphasis onto meta-systems: not the forms themselves, the details of ways of building, but the tools and rules that can be used to generate these new configurations.

New drawing instruments, geometric toys, borrowings from the developments of cartographers and navigators, the astrolabes of the Arabs, all these things enabled the creation of paper analogues of proposed building elements. These could be rescaled and transcribed into patterns on the wide floors of lofts where the huge windows and vaults were prefabricated.

Other later medieval pamphlets vary Honnecourt's magpie style, techniques adjacent to one another with no theoretical framework to underpin and organise the whole. A booklet by Matthias Roriczer (1426–1503), master builder at Regensburg cathedral, entitled *Geometria Deutsche* places pure geometric figures next to diagrams of architectural design and construction as if there were no difference and yet no connection between them. Another of his writings, *Buchlein von der Fialen Gerechtigkeit* (Booklet on the Correct Design of Pinnacles), feels modern, treating just one building element in depth, and Roriczer claimed in it to have made a direct transcription of the oral tradition of the Parler family of master masons in Prague. The outcome of his exclusiveness is also rather modern, composition as a conglomeration of well-understood set-pieces. He describes a process of design in stages. First comes a geometric abstraction, a figure drawn on paper or inscribed in a gypsum wash which only in its completeness would be transposed into a template, then finally into stone. Despite avowing only to give a 'rational' way of dimensioning spires and pinnacles, Roriczer can't resist an aside demonstrating how parts can be derived or generated in the same way as a whole has been. This bolt-on of the famous method of 'quadrature' short-circuits the emergence of a more modern structural design theory.[18]

At the same time as this groping towards more rational design and uncertain adoptions of geometry there were successive resurgences of pagan precedent. In the instruction book that the German mason Lorenz Lechler (1456–1495) wrote for his son Moritz, *Unterweisung*, practical and abstract shade in and out of precedence as the exposition proceeds. There is the reappearance of a classical proportioning system for a church structure: the unit of measure is defined by the choir wall thickness, a central structural element. The sensitivity of and variation in such issues as foundation conditions or masonry strength had not proved critical. The continual re-appearance of the numerical ratio of five to seven in the manual has little to do with geometry and more with a mysticism generated by the fascination of irrational numbers.

These mixtures of beliefs and practices achieved some of the most fabulous structures of all time. They still allow us modern engineers the preference as to how much we rely on theory and how much on craft. Since many of the rules set out in medieval pamphlets and guides had no recognisable theoretical basis, they had the potential to change with time and usage. This mutability of craft rules remains a pole opposite which modern theory sets its generalisations.

## Medieval patronage

Another variant of theory and practice's separation and re-connection stemmed from the relationship of master mason to ecclesiastical patron. Their social differentiation produced a layering of the work. The learned client acted as a systematic control, utilising the potential of building fabric to convey idea. As well as explicit iconography, symbols and cycles of narrative in stone, subtler attitudes expressed themselves. Cistercian monasteries, rigorously proscribed in their construction detail, perfectly reflect the comfortable austerity of the order. Naturally builders sought to adjust their social standing upwards. Their own strands of education, pagan arcana, reappeared as an alternative generating system for their designs.

## Animism

Alongside geometric theory in the medieval building project sat an animism well-imagined in William Golding's novel *The Spire*.[19] The narrator describes consciousness within the edifice and its surroundings. The deep structures generating the forms of cathedrals were indeed not geometric but rather anthropomorphic and theological. Different structural properties were explained with the quaintest of systems. The various timbers, hard or elastic, came from trees which mixed nature's humours in different proportions. Does this process of placing the body within the object only amount to the close observation used by builders to achieve a visceral understanding of their object? It can be detected in the numerous examples of additions and alterations made to medieval structures, not embellishment but to resolve local malfunctions as if the building's posture were being corrected. The picturesque strainer arches under the crossing of Wells cathedral (added 1315–1322) attempt to alleviate differential settlement. These interventions display a comprehension of load-sharing but a misunderstanding of gross movement. The bracing redistributes load as a muscle group would adjust a limb's alignment but the overall rotation in space induced by the set-up is overlooked. Nearby a buttress lances across a pattern of walling, showing how the masons were thinking of their load paths working independently beneath decoration supposedly applied to display structural mechanism. The Gothic of the Bohemian lands is characterised by the reduction of structure, ribs and groin lines, to incised surface ornament.[20]

## Paracelsus

Intuition promoted itself strongly through many discourses. The monk and mystic Paracelsus (1493–1541) is chiefly remembered for anticipating modern medical practice. He also left us with the notion of handicraft as the highest human expression.[21] In his writings a fierce intellectualism co-existed with incoherent alchemical ideas. As a natural itinerant he fell out with much of the academic establishment of his day and reacted by uniting a wide band of followers under a programme that elevated skilled labour and practical common sense over the abstract orderings of scholastics. Paracelsus was taking a knowledge system of another origin and recasting it for social purpose.

## Medieval shipbuilding

While intellectual tools were exchanged between Christian and Muslim, confrontations on the waters of the Mediterranean provoked an extended arms race. Such competitions tend to proceed by steady technical honing rather more than by technical breakthrough and create a milieu in which technical awareness is heightened. Muslim vessels regularly out-paced and out-manoeuvred western ships. A response was to build larger and heavier fleets acting in close proximity for mutual protection. Grapple and capture superseded ram and destroy. By the late fifteen century Byzantine galleys were heavy enough to carry real ordnance.[22] Sea battles switched from dogfights of ramming manoeuvres to exchanges between heavily loaded transports delivering marines at considerable distances from home. Hull forms combining carrying capacity and speed rather than open-ocean seaworthiness developed rapidly. Joinery and pre-stressing technology improved to meet the need, a steady refinement of lightweight timber construction. Meanwhile, along the Atlantic coasts shipping was shaped for improved carrying capacity in deep water. The cogs and bustles of the Hanseatic league, a successful trade federation of Germanic ports, were simple forms compared to the southern greyhounds but structured with a robustness to resist the wave patterns of the open ocean. Their construction relied on hull joints worked up to sustain high stresses and remain stiff under cyclic loading. Better sail-handling led to bigger rigs, increased top weights and complexity in the mast assemblies.[23] The loads generated by these pre-stressed frames, essential to driving the broad-beamed vessels over the open ocean, required reliable and 'weight-efficient' joints working over ever-extending periods of time in all climates. These types of ship would be used to undertake the European voyages of discovery. Across the fifteenth and sixteenth century, first the Portuguese then the Spanish opened deep-water trade routes towards their 'Orient' and to the New World.

There followed a particular process by which the Mediterranean trading basin and land links eastward came to be eclipsed by a much bigger ocean economy. The juxtaposition of growth and decay had a most profound influence on the direction subsequently taken by technology. The new maritime empires had novel construction problems to confront and the wealth to solve them. Conditions of slow decline in the old centres also spurred less brash initiatives to arrest or reverse the inevitable with ever more refined technologies, almost elegiac uses of cultural artefacts.

## The Mediterranean pool of ideas

The Alps and Pyrenees divide Europe, always channelling flows of people and ideas, separating environments where different outlooks might flourish then re-connect. The Mediterranean pastes together different cultures.[24] The peculiar geography of Moorish Spain linking Atlantic and Mediterranean joined two independent ship-building traditions specialised for different environments. Incessant raiding and piracy along the north African coast had refined a special kind of inshore craft. The xebec proved to be one of the fastest and most seaworthy of sailing vessels.[25] The long sturdy hull carrying large sail areas was structured with a deep keel piece. This back-bone extended into flared stem and wide counter over the stern to give good sea holding. Tall masts were stepped into the piece for strength and the additional hull stiffness allowed the boat to be driven hard. Exposed below the hull surface, this structural element also acted as a fin enabling the ship to sail closer to the wind, the key performance requirement for a pursuit craft. Different suites of sails were carried for the different points of sailing, with masts and yards extensively adjustable to meet various conditions. These corsairs ranged throughout the Mediterranean and as far up the Atlantic seaboards as south-west England. Initially adopting the type as an effective countermeasure, north European, Scandinavian and Russian navies went on to absorb its design features into indigenous forms.

Technology transfers and transformations across the north–south/east–west interfaces were part of much broader exchanges. Cross-fertilisations everywhere between Sicily and Andalusia extended from cuisine to carpentry. Many of the six-thousand-odd Arabic loan words found in Spanish refer to technical innovations and emerging concepts, particularly in irrigation and mathematics. Individuals born or raised in this milieu proved to be the source for a belief that for modern western engineers is a fundamental operating requirement.

## Lullism

Ramon Llull (1232–1315), born on Mallorca, was a restless collector of ideas, beset by various currents of belief and simultaneously isolated from the dominant schools of philosophy. His brand of theology, Lullism, is important as a very early synthesis of the idea of a universal scientific method, that the world could be fully understood and that it was admissible for man to operate on it despite his intellectual limitations.[26] This confidence and self-reliance has a complicated construction. Separating itself from the rigid systems of the scholastics, it nevertheless relies on the reassurance of universally accepted norms. These inevitable forms were still to be discovered or revealed, not invented.

As artistic, scientific and technical insights sparked along the cultural margins of the late medieval world, they are characterised as much by the direction of their momentum and continuity as by their originality. Rather than a rediscovery or recovery of classical learning coming westwards from the Arab world, it now appears more as if the Muslim understanding of classical precepts was used in the West to re-centre a knowledge base that had been continuously present through the so-called Dark Ages. It was the resuscitation of a continuous tradition of classicism, always there in the background of scholastic writings and metaphysical speculations, that led to the Renaissance. The meaning of this term and its application is much debated, even to the point of questioning the very existence of such a phenomenon. What can be said is that, commencing within the literary traditions of late fourteenth-century Italy, a distinct emphasis on enquiry permeated art, architecture, and the political and physical sciences. Undoubtedly new building techniques, a new concept of design and new training systems appeared in response to cultural change at this time. Reformation ideas that man could find or create within himself conditions for his salvation found a counterpart in a new confidence in exercising dominion over nature.

## The dome of Florence Cathedral

The dome over the cathedral of Santa Maria del Fiore in Florence is a celebrated turning point in Renaissance structural engineering (**figure 4.5**). Set upon the medieval church's crossing, its design stands at a crossroads between Gothic and Renaissance sensibilities, difficult to unravel. The construction is Vitruvian: the classical had always shimmered in and out of the Gothic, especially in northern Europe, and it relied on a very precocious engineering understanding to have been built at all.

The warring city states of central Italy, their concentrations of wealth and systems of patronage, were a forcing house for ambitious building projects. The cathedral was started in 1296 and the dome completed in 1436. That it had been laid out at such scale with such little notion of what the main cupola would be like or how it would be made demonstrates a breathtaking confidence in technology's continued expansion.

The design of the dome was commissioned separately from that of the main cathedral. In composition and structural action it was to be a self-contained set piece. Early Renaissance buildings were almost modular with their additions and re-modellings. A competition was held for the design, and the winning proposal of Filippo Brunelleschi (1377–1446) focused on issues of construction: how the dome could be made. A consummate politician in an environment rife with manipulation, he banked on a 'secret' method to raise the cupola without any centring, temporary formwork. As well as a cheap price tag, his scheme had the added attraction of limiting the time during which the crossing area would be filled with scaffold and otherwise unusable.[27]

How did Brunelleschi synthesis his 'unique' solution? Trade and cultural investigation had led him in his youth eastwards, where he would have seen the ruins of Byzantium. The old domes decayed in a specific way (**figure 4.6**); rain and interstitial condensation from within rotted out the crowns of the domes until partial collapse left most of their shells intact but open to the sky. Brunelleschi realised that if such dilapidated structures held up then by reversing the process in the mind's eye one could see a dome gradually being filled in towards its centre, all the time

4.5
**Santa Maria, Florence, 1296–1436**
The competition-winning design for the dome of Florence cathedral is idiosyncratic. Construction method became a central issue in the straitened circumstances of the city at war. Brunelleschi's proposal for an octagonal ribbed structure required no centering. The ribs supported a substantial lantern without undue spreading.

**4.6**
**Hadrian's Villa thermal baths, Tivoli, AD 120**
Brunelleschi advertised his brilliant insight and was allowed to proceed without revealing how he would avoid temporary works. Having travelled widely in the east, he must have seen many ruined domes open to the sky. If they remained stable as their crowns eroded then the process must be reversible.

remaining stable. A cupola could be built up as a series of rings closing towards the middle without centring. The bricklayers would work from outside, leaning over the decreasing hole within. He further added to his design a double-skinned structure, lightweight and shaped to contain the thrust lines involved. A pointed profile could carry a heavy masonry lantern and also reduced hoop tensions (spreading), running round the bottom part to a level that could be dealt with by a cramped ring of stone. Close observation and the isolation, then re-combination, of distinct structural mechanisms gave him a structure of immense sophistication and verve.

Large-scale timber models aided the conceptual processes of designing. Repeated presentation of the scheme by designer to client was an intrinsic part of its development as the ideas used had to relate to the understanding of a lay audience of commissioners. Such occasions seem to have fostered two approaches: either the concept was simple enough to explain and grasp easily, or it had to be of extreme complexity, veiled enough to require an act of faith. Modern structural design retains something of this underdeveloped level of communication.

The histories and descriptions of Brunelleschi's building project all reflect another transformation under way, again a scale effect. Material handling became a major influence on the details of the design and a new generation of cranes were developed as part of the work. These temporary mechanisms, derricks and booms, had to be safe and very efficient.[28] Their configurations magnified already large loads on the bearings of their pivots and pulleys. Extrapolating from the contingent construction of siege engines, the continual redeployment of craneage allowed load paths and strain patterns to be observed and amended. The construction site itself became an environment for structure to evolve in.

## Leonardo da Vinci

The paradigm of Renaissance man is Leonardo da Vinci (1452–1519). His continual observations, notebooks full of machines, dissections, natural phenomena and processes are all directed to the mastery of nature through its perfect depiction. Science was an aid to the painter in his recreation the world. Implicit is the notion that there is an underlying coherence and reason to the universe, a consistency to be tapped into. The apparent modernity of his ideas veneer a very different consciousness, a method much removed from our own.[29] Leonardo applied contemporary humanist models which placed man as the measure of all things. Everything was literally anthropomorphic, from geology upwards. His work mirrors the sensibilities of his time, and our re-interpretations to find new significance for him reflect our own.

Leonardo's studies for new bridge designs display only a partial understanding of the mechanisms involved (**figure 4.7**). His siege bow in the Atlantic Codex simply would not have worked. Partly, this may be to do with imprecise copying from the work of others – not necessarily a bad thing (misunderstandings and mistakes can generate new insights) – and partly due to his genuinely creative technique of suspending some elements of reality to see through a thought experiment.

The celebrated drawings of machines mix realistic detail with flights of fancy. Large mechanisms are in timber, junctions pegged and banded with iron to sustain the real load concentrations that would have occurred. Many of the structures proposed would have been compromised by joint slippages and excessive deflections. The flying machines were far too heavy.

The Renaissance had many permutations and shadings both sides of the Alps and in between the Italian states. It was to be in the unique set of conditions pertaining to the Venetian Republic that modern structural engineering became fixed. Although the city had thrived in the crucible of Mediterranean conflicts, trading and allying itself to create an empire, it was its protracted decline that made it a wellspring of modernity. Renaissance fine art was reformulated in a distinct idiom by Jacopo Bellini (1400–1470) and his sons Gentile (1429–1507) and Giovanni (1430–1516), the latter's experiments with oil-based paints using the exotic colours available from the

4.7 (*opposite*)
**Codex Atlanticus, Leonardo da Vinci, 1478–1519**
The sketches and notebooks occasionally contain the slips of a copyist, potential sites for inspiration. Charged with developing siege engines, Leonardo explored an extensible bridge in this drawing. An elementary structural form repeats to form a cantilever. In practice, the multiplying bays would quickly become too flexible.

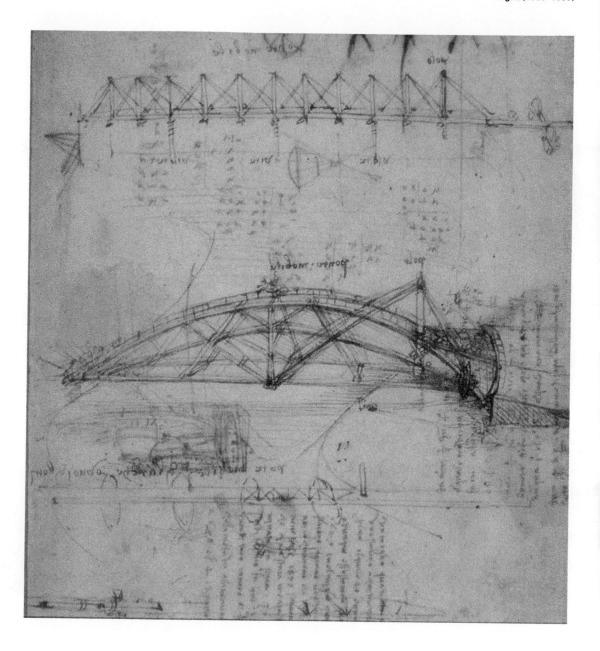

East released an unprecedented realism.[30] This technological entrepôt extended into building forms and naval architecture. Always strapped for manpower and materials, the militarised society came to rely on better technology and manufacturing techniques for its edge in war. The turning point came with the return of the Portuguese explorer Vasco da Gama (1460–1524) to Lisbon in 1499, having found the sea route to India. From then on Venetian trading pre-eminence began to dwindle. The distillation of knowledge combined with relentless pressure for practical solutions concentrated the development whose outcome is our modern way.

# Chapter 5

# Galileo Galilei (1564–1642)

Modern structural engineering is presented to its students as a system, complete in all its parts, applicable anywhere, for all conditions. Quite how we've become entranced by such faith has already been touched upon. Later chapters will review whether anything like its realisation has been achieved. However close we have now come to a universal method, it is one still heavily marked by the extended process of reaching it, a process based on an accretion of ideas, assembled around contingencies, situations demanding immediate and pragmatic solutions, so-called 'quick fixes'. And yet despite this piecemeal approach, the essential shape of structural engineering was definitively moulded in a very short space of time, by a few individuals, all working in the vast sea-change of consciousness labelled the later Renaissance.[1] Various histories have it that a modern engineering consciousness sprang up almost fully formed overnight to be rapidly accepted by the establishment, and indeed the predominance of an 'Italian model' merits explanation. Perhaps a new approach to structure was a delayed inevitability, over-determined by historic events and environment. That underplays the lost value of alternatives that were subsumed but echoes of which might still be detected to broaden our discourse.

## Venice

The big change originated in early seventeenth-century Venice. As the city state declined, retrenching between more powerful neighbours, she compensated with a brilliant outpouring of art, architecture and literature. Peculiar circumstances afflicted her economy, extreme wealth coupled with a shortage of the manpower needed to maintain continual and far-reaching military effort. As well as warring, there was endemic piracy to be suppressed and money to be made from mercenary soldiering. Good equipment would make up for the lack of numbers, and desperation and trading acumen were prepared to pay the high price of technological advance. The good years had established a strong industrial/military complex. Social conditions had

5.0 (*opposite*)
**Boat-building on the beach, Greece, today**
Boat-building on the shore of the Mediterranean. The lines of the hull are set out by the keel and cross members. Once complete, frames and planking make a shell stiff enough to span the waves, but it is in this construction phase that the ribs are most heavily loaded.

decided the issue of quality over quantity and skills had concentrated in the islands. By the fifteen century deforestation was already well advanced down the coast of Dalmatia and the premium on timber, work space, even water for manufacture, in Venice promoted a consuming care of and economy in material use. The galleys and cargo ships of the republic were efficient structures.[2] The sheltered conditions in the lagoon offered an ideal environment for the weapons testing and trials essential to successful and safe innovation.

## The courtesan Galileo

Galileo Galilei became involved with these efforts. Born into a wealthy manufacturing family in Pisa, he neutralised weak social credentials by inventing a career trajectory through the courts of Italy as a scientist. He used experimental research, its presentation and dedication to benefactors, as a tool for advancement.[3] These machinations as a courtier were finally to catch up with and consume him when he fell foul of a Dominican pressure group within the Catholic Church. Forced to recant his support for the Copernican revolution (though heliocentrism was never a heresy), he spent his last eight years confined by the Papal courts to a villa at Arcerti. This house arrest he used to recapitulate many of his scientific discoveries, and to extend his ideas in the famous book *Two New Sciences*. Subsequent developments have made this work recognisable as the first modern structural engineering treatise. Galileo's approach was very much based on a reconciliation of two influences: one the procurement and manufacturing base of the Venetian Republic, the other his own social and political milieu.

## Standardised ship designs

The Venice Arsenal produced its war galleys on a production line, the first factory capable of making all the components of these complicated artefacts in one place. Designated areas within the vast site contributed quality-assured products. An intricate division of state and private initiative was given responsibility for ship-building so as to encourage innovation and development. Upon launching, a new hull would be towed along a canal and fitted out systematically from quayside stores as it proceeded; women handed out fittings and rigging from first-floor windows to finish a seaworthy vessel in the shortest possible time.[4] Everything was compartmentalised.

It was imperative to recycle war- and storm-damaged ships wherever possible. The first fully interchangeable parts were hung rudders.[5] Stern areas were vulnerable to the waves and to attack, so these components were the first to be lost. Instead of being made to measure for each sternpost, rudders were standardised and mass-produced. One size fitted all. It became worth the time to refine the design. In the ramming battles of the time, hulls were stoved in and cross frames broken; damaged prizes and allies were regularly towed off for reconditioning. Regularised rib profiles gained acceptance with shipwrights, and were prefabricated then held in reserve for speedy repairs.

## Abstracting an engineering problem

Entering upon this enterprise, Galileo chose to focus his contribution on the detailed study of one particular problem. Narrowing bandwidth to make information manageable is a characteristically modern technique and the losses thereby should be acknowledged. He came to a consideration of the strength of materials with a simple conception of stress; he had worked through the idea of a 'scale effect' – that for a given strength of constituent material, objects can only grow so big before they collapse under their own weight. This insight he based on the proportional arguments, the classical method that he so often relied upon.

Reducing the hull weight of vessels without sacrificing strength was a problem apparently quite beyond any rationale of the time. Venetian shipwrights kept the ideal proportions of their designs partly encoded in song and they would effectively sing out their dimensions, just as Figaro imposes harmony on his domestic surroundings. Despite the ferocity of Adriatic storms and the jarring impacts of battle, the strength of ships' rib frames proved to be most critical during the shipbuilding phase. Construction commenced with the laying down of a keel piece and ribs were then set out along each side and mortised into place; the skeleton was planked in and the hull decked over. The cross frames were most highly stressed before the shell was complete. If they were propped then the shores had to be continually relocated as the work proceeded, delaying progress; otherwise the ribs had to act as cantilevers carrying the entire weight of the hull until it became an integral whole. Shipwrights were over-sizing elements just to meet a temporary condition.

Galileo focused on this relatively tractable problem by seeking to find the minimum safe size for rib cross sections – that he and others expected a rational solution to be achievable at all is telling. He used abstraction, reducing the problem to its essentials, and in the process made its solution applicable back onto a much wider range of problems; (**figure 5.1**) the method he used is an odd concoction, mixing Archimedes' lever principle with the newer notions of algebra. Crossing from the Arabic world to the West through trading portals such as Venice, the manipulation of symbols had by then transformed mathematical speculation. Simultaneously introduced, both algebra and the classical mechanics recovered from the East became coupled together. Some of Archimedes' personal attitudes also came across; despite an astonishing range of practical invention, he always held that pure mathematics was the only worthwhile pursuit. The work of the Syracuse engineer had been retrieved through the re-interpretations of Pseudo-Archimedes, published in bad translation by Niccolo Tartaglia (1500–1557) in 1543 and then in accurate mathematical treatise form by Federico Commandino in 1558. Literature of the time displays the common intention of making mechanics and engineering the true goal of mathematics; the building of machines was to be a mathematical art, with everything reduced to the theory of the lever.

## The reception of engineering ideas

Galileo's analysis of the bending of a cantilever turned out to be wrong;[6] his method was flawed and overestimates the strength of a section by some 50 per cent. With

5.1

**Galileo's abstraction of a cantilever, from Dialogues Concerning Two New Sciences, 1638**

The problem of sizing hull ribs to be just strong enough during construction was reduced by Galileo Galilei to the consideration of an abstract cantilever. His solution imagined the structure composed of Archimedean levers. With no understanding of elasticity, he failed to see a more realistic mechanism.

no concept of a relation between stress and strain, he overlooked the way timber crushes at nearly the same stress at which it tears. The additional abstraction of a 'neutral axis' was needed to make his model work, and the groping about to find that idea took another forty years. What is interesting, however, is that his method was not simply rejected but found use: a soon-to-be-superseded 'pragmatism of science' then prevalent saved it.[7] If it was accepted that man could only have an imperfect idea of God's creation, a mystery only partially revealed, then it was always good enough merely to 'save the appearances'. Whatever worked better was to be accepted: and so rationality was not the test, just the nearness to theological truth.[8] Neither Galileo nor his jailors thought him to have been imprisoned to suppress a truth but rather over an altercation as to what that truth might be.

Setting up mathematical models to represent real situations, then extracting insights and generalisations from them, is modern engineering's staple. In all his work Galileo was solving problems for things already in existence – war galleys were well dimensioned long before his analysis of them. This relationship characterises the method and persists today: given form is the impediment of modern structural engineering and is also fundamental to its make-up.

What was the 'acceptance' of Galileo's ideas and how did it become so complete, established and influential on what follows? The process has been represented elsewhere as a Thomian 'paradigm shift', an over-determined inevitability.[9] The times fostered his activity and he filled a gap. It might be better to recast the explanation admitting a little more contingency, rather more as being a historical event. The background to the work was undoubtedly important. His output coincided with fundamental changes in ways of communication and his approach relied on a much wider information revolution emanating from the Renaissance. Experimenting with cut-down wine presses in the Rhine valley, Johannes Gutenberg (1397–1468) had revolutionised printing, and hence literacy, with his invention of movable type; the first book produced by such means, the Mainz Bible, appeared in 1440.

Rather than being a trigger to an immediate shift in the perception of information, this invention instituted instead the slow pervasion of new power structures within society. Its exploitation was steady rather than explosive, corresponding to and influencing, but also perhaps mapping, a 'background radiation' of intellectual expansion in the Renaissance.[10] Closely linked with the trading world's publishing power, a commonwealth of education meant that individual idea sets could dominate and extinguish others. The dissemination of technical ideas as they occurred would be rapid. Each language group would become a complete world; no longer would there be conglomerations of disparate conceptual spaces. Galilieo's *Two New Sciences* was printed and published by Elsevier of Leiden in 1638.

In Italy those interested in the new learning began to group together, establishing peer groups that could independently determine what the acceptable canon should be. This movement started in Naples with the foundation of the Accademia Secretorum Naturae in 1560. The Accademia dei Lincei was founded in Rome in 1603, with Galileo as one of its members.

## Simon Stevin

The Pisan researcher had irrevocably linked science and mathematics in a particular way. Elsewhere confrontations with practical problems were being met with other forms of quantitative abstraction. A typical example of an engineer straddling the medieval and modern divide was Simon Stevin of Bruges (1548–1620), an intensely practical man who started out as a bookkeeper. The neglect of geometry in favour of arithmetic prevalent in the Netherlands at the time favoured the practical approach he took to engineering. Stevin is famous for defending the nascent Dutch republic by deliberately flooding large areas of polder to keep the Spanish at bay – a kind of Dutch Barnes Wallis making bold manoeuvres in the desperate days of the revolt. It is his appropriation and transformation to something practical of the new mathematics that is of interest here.

Stevin's ideas centred on statics, notions of bodies at rest – a preoccupation that has remained with us, detrimental in that it sets aside dynamic treatments. He argued from the impossibility of perpetual motion towards a concept of equilibrium.[11] He followed a logical quasi-mathematical approach, proposing thought experiments, imagining structural behaviour and seeing through the consequences with numerical analysis. He was aware of more rigorous mathematical expositions – books such as *Mechanics*, published in 1577 by Commandino's pupil Gidobaldo del Monte (1545–1607) – but used their results sparingly and in piecemeal fashion. The demonstration of results by appeals to common sense rather than mathematical rigour is another legacy of his to modern engineers.

The Dutchman published his own writings, first in Flemish, from 1586 onwards, then in Latin editions between 1605 and 1608, and eventually in French in 1634. Despite this effort, problems of accessibility meant that the formulations of others largely overtook his ideas. His systematisation of the parallelogram of forces, however, has become the standard model we use today. In creating that diagram he probably took the notion of a parallelogram of velocity raised by Archimedes and recorded by Leonardo da Vinci and then 'froze' the dynamics. This crossover from geometry to graphical statics imitates the mason's embodiment of forces in the stone, channelling and deflecting vectors down across the faces of their buildings.

## The calculus

Stevin's writings perpetuate some concepts with very old roots that later came to be fundamental to the mathematics of modern engineering. The manipulation of water, embankments, dams and floating barges was his main concern. In order to help himself with hydraulic pressure problems and to find the centres of gravity of complicated forms, he invented a method presaging modern calculus.

The development of calculus, the maths necessary to model change and control in nature, had a long gestation before the key concept of a function approaching a limit was identified.[12] Its origins lay in the ancient Greek method of exhaustion, a geometric process invented by Antiphon the Sophist around 430 BC in which a shape would be described by increasingly detailed approximations until the difference became 'exhausted'. The idea came to Stevin through the applications made and recorded by Archimedes. The Renaissance engineer was using the conceptual tools transmitted to him and applying them directly to the problems before him. Their successful solution seems to have stopped him taking the additional intellectual step to create the concept of limit and therefore unlock the general potential of the mathematics. This had to wait another century. Stevin left his method as just another tool in the bag.

## The influence of astronomy

Mathematicians and artists were establishing new ways of seeing. The astronomer Johannes Kepler (1571–1630), a German sponsored by the Danish court, proposed the infinite point for use with conic sections. The device was applied to generate linear perspective by Girard Desargues (1591–1661),[13] who wrote densely worded theoretical treatises on stereotomy, the cutting of stones, and on sundials as part of his development of projective geometry. The Frenchman almost seemed to be trying to uncouple his work from its practical origins. The need for mediation in order to truly see, anamorphosis, seems to have been universally accepted.[14] Previously perspective had been precluded by the breakdown of temporal continuity. Abstract, static situations could now be conceived of and explored.

Throughout this period astronomy held onto its reputation as the best route to the understanding of the world and so came to colour the mathematics passed down to us. Much mathematical invention and discovery, such as John Napier's (1550–1617) algorithms, was in response to the immediate requirements of observational astronomy and the changing cosmology. Much of modern mathematics as it has developed is an indirect reflection of the heavens which in turn influence engineering by that indirect route.

Music was another governing influence on mathematical development with its structuring by proportion. The mason's rules based on practical sizes were supplanted in a similarly manipulable format by divisions based on harmonies.[15] These ways of spatial thinking were relevant to the changes that took place in the mathematical forms used to represent nature. Signs gave way to spaces. Stable and unstable oscillations were detected all around.

At the same time that astronomical research, musical and graphical theories were shaping mathematics so also the protracted warring between the city states of northern Italy gave a special impetus and emphasis to the development of geometry and construction. The lack of manpower and inventions arriving from the Near and Far East rapidly escalated weapons improvement, particularly of artillery and firearms. Sieges involving resistance to heavy bombardment became the norm and castle-building responded with new substance and geometries.[16] Trajectories of fire and fields of view dominated the setting-out of plans; surveying techniques were devised and improved to cope. Just as perspective was being refined for pictorial representation, so analytical geometry was being developed to fully describe the spaces in and around the new emplacements.

Mental models of the universe underwent a radical change of emphasis. Always having been designed to explain the appearances, they were now proving liable to rapid supersession. It was also becoming obvious that the real world must be tractable and available to ever more detailed and sophisticated systematised thought.

## Algebra

Among Galileo Galilei's many correspondents was a Croatian scientist and mathematician, Marin Getaldic (1568–1626), whose work was to contribute as much as the Italian's to the nature of modern engineering. Through his work on optics he set the grounds for the algebraicisation of geometry.[17]

The special impetus given to the development of algebra in the Arab world stemmed from Islam's complex inheritance laws and the need to work through large family divisions. Transmitted to the West, the subject was not initially developed in its own terms but applied back onto the geometric axioms. Getaldic solved forty-two specific geometric problems by manipulating symbols and the process was generalised and carried to its present form in the work of René Descartes (1596–1650) and later Pierre Fermat (1601–1665). They established a solution procedure to geometric problems using algorithms, rules rather than imagination; discoveries could be made within the numbers rather than invented in space. Several of the more complicated algorithms were set up at the time and the manipulation of abstract quantities became a commonplace. For some commentators Descartes' contribution has been identified as the end point of a development rather than the revolutionary initiation of something new.

By means of this algebraicisation of geometry, many more surfaces could be described and mathematical modelling possibilities ramified than could be drawn. New surfaces could be invented, or rather found. If it were accepted that reality had a level of complexity beyond geometric solution then algebra offered a way of dealing with it. This tension between number and space epitomises the consciousness of the time and is not yet discharged.

These changes seemed as revolutionary then as now. By any definition the Renaissance ushered in a new age. But there were other parallel understandings, self-organising into bodies of thought. That they were superseded may not have much to do with any intrinsic inadequacies.

## Mysticism maintained

Of the alternative thought systems, one of the more developed was Rosicrucianism. The mystery surrounding this movement was partly self-generated and partly a consequence of its gradual supersession.[18] Traces of its intellectual framework can be found in our modern consciousness, transported there through the work of Descartes, Bacon, Kepler, Newton and others. Magic, alchemy and the cabala were parts of it. The English polymath John Dee (1527–1608) is representative of its shadowy membership.[19] He worked in the mainstream and margins. As well as for contributions to medicine, he became celebrated for his demonstrations, visual tricks and experiments. A central tenet of his was that this world consists only of appearances, Paracelsus' 'solidified smoke'; it is a system of signs, which can be read and manipulated but which also veil a deeper reality. Such notions were prophetic; there were indirect ways of controlling nature. Base metal can now be engineered into other elements, and our perceptions altered and corrected.

The practitioners of these 'arts' looked to science and mathematics as the way to 'unlock' deeper truths. Rather than developing practical mental artefacts they were seeking portals to other actualities; their maths and physics could be used to research the nature of reality, a new metaphysics. Such studies contained within themselves the knowledge that was being sought independent of external phenomena. The idea that mathematics is arcane, that it embodies absolutes, is still with us, as is the acceptance that engineering can reach out beyond the apparent to the seemingly impossible. Science itself had its own internal logic which made it more than a set of inventions; it was a complete alternative world within itself.

## Machines

These thinkers found it fashionable to maintain a parallel interest in machines, clockworks and automata, as had Vitruvius. The development of the required precision of manufacture was self-sustaining. Steel screws first appeared to fasten plate armour, then to secure the locks of arquebuses despite repeated firings,[20] and miniaturised screws began to be used in clocks from 1550 onwards. Lathe work in metals was transformed in quality and accessibility; tools that could improve and economise gear manufacture, blocks and pulleys, required better materials and supported the artisan base of the thriving towns. Machines were made stronger internally to improve their reliability, and size and weight could be correspondingly reduced. Improving accuracies in measurement and instrumentation attended these innovations.

Political renewals across Europe resuscitated Roman law. The ancient codes of the Emperor Justinian had the effect of separating areas of action from the direct control of religion, thereby encouraging a breadth of thought and more progressive attitudes. The stultifying over-critical intelligence of Greek thought was also being set aside. The Italian theologian Thomas Aquinas (1225–1274) laboured to subsume Aristotelian metaphysics to Catholic dogma,[21] and this ensured its rejection by many Protestant thinkers who turned to contemporary resources for their ideas.

## The method of Francis Bacon

Francis Bacon (1561–1626), courtier to James I of England, worked through what this new rationalism, which took on a subversive role in drama, politics and society,[22] might mean. Bacon was, by all accounts, a difficult man but one who wanted social change and who believed technology could and would bring it about. The universe was a problem to be solved, examined and meditated upon, rather than an eternally fixed stage upon which man walked. This recognition, together with an unlimited confidence in its potential, made him a tireless, erudite advocate and author.

His dismissal from legal office for taking bribes left Bacon with a cause to prove and he set out to show that technology could transform man's limitations. By the use of practical devices humanity would be able to perceive and project more clearly. Social systems would be transformed; the great systems controlling our thoughts (such as Aristotelianism) would be extinguished; problems of language and communication would evaporate.

Bacon recognised that many inventions would come from technical and intellectual crossovers. He differentiated between science-based findings and empirical inventions, and his preoccupation with the significance of the great geographical discoveries of the time led him to rather overlook the mechanical arts and the coming importance of mathematics. Trained as a lawyer, he came up with a version of scientific induction in which a succession of particular findings is built into a general principle. This method has had its critics (how could a general proposition ever be arrived at conclusively?), but though rejected by scientists keeps its usefulness for engineers. Despite its flaws, the work of Francis Bacon became a rallying point for other generations of dissenters around whom the Industrial

Revolution in England took place. His gift to science of an independence from religion and government similar to that enjoyed by judicial law must be priceless.

## Andrea Palladio's generating system

Some of the ways in which engineering was directly influenced by classical learning in the Renaissance are illustrated in the research undertaken by Andrea Palladio (1508–1580). Born in Padua, a possession of the Venetian Republic, he was apprenticed as a stonemason. Close study of recovered texts, particularly Vitruvius, and the ancient monuments led him to create an extraordinarily influential architecture – for which he found a sophisticated, receptive market, designing country villas and urban palaces for the successful landowners and traders of the Veneto. Tasked with the construction of several small and medium-sized bridges across the embanked rivers and drainage canals of the area, Palladio searched the literary and pictorial, but seemingly not the archaeological, records for classical precedents, lighting on the celebrated bridges: Julius Caesar's across the Rhine and Trajan's across the Danube (see pages 26 and 29).[23] These examples he promoted as perfect – quite why, given their contingent nature, is not clear.

His design process was to study, then interpret, their construction for reuse (claiming accuracy by corroborating between sources). It little mattered that these structures might have been for other uses, such as military application across very wide rivers, and made with more primitive tools, or that their design was being transmitted with all the distortions of the literary forms, the sculptors' conventions of relief-carving and the inescapable influence of contemporary knowledge. Palladio believed that somehow a timeless element, knowledge as a deep form, was being transmitted.

Despite his declared intentions, his reinvention of a variety of timber truss structures reads rather more into the record than was there – another example of engineering anamnesis,[24] a recall of form already embedded in the mind. There is more than book learning here. The reliance on the authority of ancient authors is underpinned by an intense effort to see order and effect in the universe. This deeply held desire to subdue nature would subsequently inform the Baroque spirit just then ascending. As it turned out, the classical remit offered just the right bandwidth of vocabulary to produce a thoroughly modern series of forms within the old syntax. The Palladian bridges are indeed a complete exploration not only of classical precedent but also of the contemporary level of structural understanding. The trusses are a set of elementary sub-systems superimposed on one another or piled up in hierarchy, all rigorously controlled by proportional rules (**figure 5.2**).

## St Peter's, Rome

The dome of St Peter's, completed in 1624 to the designs of Michelangelo Buonarroti (1495–1564), is a product of Gothic practice fully infiltrated by Renaissance ideals.[25] Directly comparable in size to the crossing of Florence cathedral, the design no longer allowed construction rationale to precede architectural effect (**figure 5.3**). The shell's

5.2
**Ponte Vecchio, Bassano della Grappa, 1569**
The classicist Andrea Palladio believed he was systematically recovering the knowledge of the ancients, a true and complete understanding of the world. In the absence of substantial archaeological or historical records, his bridge designs are more rational speculations than reconstructions, condensing contemporary knowledge through renaissance intuition.

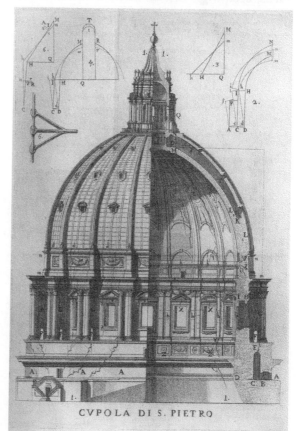

5.3
**Dome of St Peter's Rome, 1624. Analysis by the Jesuit Ruggiero Giuseppe Boscovich (1711–1787)**
The management of a structural system through systematic observation, modelling, analysis, assessment and intervention that occurred in response to cracks appearing in the original design by Michelangelo Buonarotti (1475–1564) completed by Giacomo Della Porta (1533–1602) was thoroughly ancient in conception and utterly modern in execution.

CVPOLA DI S. PIETRO

thickness was attenuated; circumferential ribs were omitted and radial ribs much reduced. The dome was set up high on a rotunda to improve its external appearance. But the structure didn't work well, beginning to crack as soon as it was constructed, and a protracted series of improvements and interventions were made to control the spread of the dome, which had been incorrectly assessed. The continuing problems became a test bed for different concepts of structural action. A new way of finding an engineering consensus was establishing itself, one characterised by a sequence of phases; from an inception, the onset of an unforeseen circumstance (divergence), there would be the testing of alternative solutions, followed by the acceptance of one best method (convergence). The conclusion of the process was a reassuring supersedure of the old inconsistencies.

## Mannerist engineering

Further appropriations of classical precedent grew into a self-conscious emphasis on technique at the expense of cohesion and content, later labelled Mannerism. In engineering, this reduced to a concentration on parts rather than wholes, contained structural devices which could then be collaged together. This conceptual partitioning of elements opened the way to the studies of individual systems, arches, vaults and beams, upon which modern theory is based. Michelangelo is now recognised as a progenitor of this approach, obsessively working over a limited repertoire. The sculptor, painter and architect is credited with an involvement in the design of the Ponte St Trinita (1567) over the Arno in Florence (**figure 5.4**), a low-arch bridge using

5.4
**Ponte Santa Trinita, Florence, 1567–1569**
Elliptical forms permeated the new dynamic universe from Kepler's planetary orbits throughout baroque design. In this project by Michelangelo, the track of an object falling to earth transposes into being the envisaged load path. Oddly, the parabolic trajectory was turned ninety degrees and tilted slightly to achieve an elliptical approximation.

a parabolic profile but inflected in an extraordinary way.[26] The curve is halved, then rotated through 90 degrees and rejoined to its other half on the 'wrong' ends; the result needs a slight visual correction, each side tilted up to iron out the peak that would otherwise occur. This profile is a step beyond the ellipse, that restless form then being discovered everywhere, in Kepler's heavens and in the ground plans and façades of Counter-Reformation churches. Beyond Mannerism, the seeds of the Baroque explosion had germinated. The spherical geometry of the navigators became available to architects, and from the Sforza Chapel, St Maria Maggiore, Rome (1564), onwards a profusion of new volumes and intersecting spaces would be dealt with.[27]

Profound economic change attended these intellectual advances. The new discoveries, the opening of the sea routes to India, now re-centred trade development in the West onto the Atlantic seaboard at the expense of the Mediterranean basin. Many of the new ideas would be taken up by Protestant sensibilities and transmuted and enriched by fresh outlooks and insights. Ships would have to carry the new wealth, and warships to fight over it; buildings would have new forms to accommodate it and new scales at which to display the new power systems.

# Chapter 6

# Early modern engineering (1580–1789)

The reception of Baroque sensibilities in Britain is difficult to interpret, complicated as it was by an attendant antipathy towards the Counter-Reformation. Nevertheless, profound changes occurred and the culture was sufficiently permeated by multi-faceted ideas to form a foundation for the onset of the 'Enlightenment'. A persistent Baroque meme[1] involves dismemberment, independent study of parts then reassembly of wholes, and in this process ambiguity and juxtapositions find room.[2] The ancients' intuitive grasp and animist feeling for whole building structures was replaced by a deeper knowledge of the parts. Overall understanding needed retrieval and structural interactions were systematised – this drawing together being character-istic of the Age of Reason. That catch-all words Enlightenment, Aufklärung, Siècle de Lumière are used here to denote the insistence on man's destiny over nature; not god and/or man but divinity in man.[3] A terrible responsibility comes with the social and technical tools to transcend one's immediate state. The single word envelops a series of parallel developments, perhaps better described as separate 'enlightenments' in a complicated period of disparate attitudes to technology.[4] A diversity of concepts and inconsistencies were in the process of resolution or replacement.

## Sir Christopher Wren

The precursors of the new sensibility moved in a charmed world. Typical of northern polymaths – at one remove from the Mediterranean but following in the traditions of Galileo and Leonardo, and who fixed some of the relationships that were to develop between the disparate enquiries of sciences, art and culture – was Sir Christopher Wren (1632–1723). In a rigidly hierarchical society, the convolutions of the English Restoration allowed him to progress from his relatively lowly beginnings, as a Dean's son, to become Controller of the Royal Works.[5] The ferment of thought and liberalism in these times manifested themselves in his coffee-house rationalism and cross-breeding of ideas. Together with his Royal Society colleagues he attempted to imbue his own brand of neoclassicism with a coherent rationale. Their endeavour

6.0a (*opposite*)
**Dome of St Paul's Cathedral, London, 1611–1711**
An intensely pragmatic response to raising a dome and lantern high above London. Without being sidetracked by recent advances in dome theory, Wren realised the central problem was to carry the 400-tonne lantern. A cheap and forgiving cone of brickwork distributes load evenly outwards. Lightweight framing and a very modern cladding system make the outer dome into a weightless stage-prop profile.

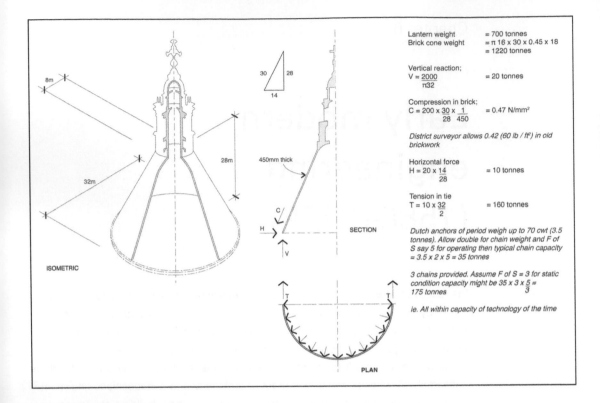

Lantern weight = 700 tonnes
Brick cone weight = π 16 x 30 x 0.45 x 18
= 1220 tonnes

Vertical reaction;
V = 2000 / π32 = 20 tonnes

Compression in brick;
C = 200 x 30/28 x 1/450 = 0.47 N/mm²

*District surveyor allows 0.42 (60 lb / ft²) in old brickwork*

Horizontal force
H = 20 x 14/28 = 10 tonnes

Tension in tie
T = 10 x 32/2 = 160 tonnes

*Dutch anchors of period weigh up to 70 cwt (3.5 tonnes). Allow double for chain weight and F of S say 5 for operating then typical chain capacity = 3.5 x 2 x 5 = 35 tonnes*

*3 chains provided. Assume F of S = 3 for static condition capacity might be 35 x 3 x 5/3 = 175 tonnes*

*ie. All within capacity of technology of the time*

ISOMETRIC
SECTION
PLAN

came with theoretical underpinnings, its consistency based on a belief in a deep structure of the universe revealed in the workings of the heavens and in human form. In Wren's building designs, plans and proportions freely combine anthropomorphism, classical principles and dimensions derived by the calculation of statics.

> Wren was an astronomer and mathematician. He enjoyed gadgets. His breadth of interest appears to have been un-compartmentalised – structures of all sizes up to that of St Paul's Cathedral in London, his culminating commission following the Great Fire of 1666, were designed within an intellectual continuum recognising no separation of ordinance and structure.[6] This was not a close-worked effort at synthesis but a deliberate attitude; the ascendancy of theory over practice was established. Wren the astronomer turned architect, and his buildings inevitably presented themselves as macrocosms.

6.0b
**Lantern of St Paul's Cathedral, London, 1706**

## John Vanbrugh

John Vanbrugh (1664–1726) collaborated with and learnt his architecture from Wren. His experience as a soldier, arrest for espionage and controversial career as a playwright all equipped him with an attitude and wit well suited to confronting a new architecture. Demand for rich houses was burgeoning, and Vanbrugh's Baroque estates were literal backdrops on which to project a new social order. On behalf of a grateful nation he built a palace for the Duke of Marlborough, Britain's hero in the War of the Spanish Succession, which had reset the European balance of power.

His pavilions and link blocks make up confections, sometimes only one room deep, stretching to embrace a surrounding abstraction of landscape.

## Nicholas Hawksmoor

Wren, then Vanbrugh in turn, were assisted by an able master builder. Nicholas Hawksmoor (1661–1736) was born into a farming family but rose rapidly through a series of 'hands-on' posts as clerk of works and surveyor. Eventually in receipt of commissions of his own, the seemingly practical mason began embedding Rosicrucian arcana into his ecclesiastical projects, colliding architectural devices together and experimenting with oddly complicated timber framing and deceptive structure, creating uneasy effects.[7] If the Baroque sensibility never really took full hold in England then Hawksmoor can be seen to have introduced a personalised alternative into that vacuum.

## Theatrical structure

If structure could be reduced to decoration then its index as cultural carrier could be completely explored. Many of these English Baroque buildings were badly built, effectively little more than extended theatre sets with stone and brick substituting for timber and tar paper. The dome of St Paul's in London is really a concealed brick cone carrying its massive lantern (**figure 6.0a** and **b**). The perceived weightlessness of the lead bubble outside is real – just a timber staging carrying a boarded profile.

## The theory of elasticity

As described in the previous chapter, Galileo's study of the beam problem went astray but the invention of a simple conceptual tool made it serviceable again. The foundations of the modern theory of elasticity condensed in the work of Robert Hooke (1635–1703), another son of a churchman and a co-founder with Wren of the Royal Society. This group self-consciously promoted the new rationalism and scientific method despite already being dimly aware of its 'internal inconsistencies'.[8] Intended as open fora of information exchange with discoveries as common property, such scientific societies were beset by personality clashes. Results could not be assessed with indifference, and competition among the researchers must not be underestimated even in this seemingly halcyon time. One fellow of the Royal Society and eventually its life president, the mathematician Isaac Newton, reminded Hooke of Bernard of Clairvaux's aphorism that 'if we see further it is only by standing on the shoulders of giants'. Robert Hooke was hunchbacked.

Hooke's range as an experimental scientist was boundless, from geology to zoology. Out of his work tearing and breaking materials he derived a simplification, now known as 'Hooke's Law', for the relationship of stress and strain in a wide variety of substances. Linear elastic behaviour is a poor shadow of the complexity of real world actions but it releases a floodgate of analytical possibilities once it is accepted. If stress and strain change in direct proportion to one another then relatively simple mathematics were already available to predict structural behaviour accurately. Structural materials began to be adopted and developed precisely because they suited this amenability. The relationship is a fair approximation to the characteristics of such stuff as ductile iron and lightly stressed timber; however, the model has been somewhat forced to accommodate concrete and masonry, far cries from the elastic ideal. Our most recent models of materials behaviour revisit some of these problems but often only within the framework of the theory of elasticity already set up.

## Simple bending theory

So the French physicist Edme Mariotte (1620–1684) unravelled the many confusions subsequent to Galileo's attempt to complete the bending equation. He is celebrated as the originator of the spirit of doubt and timidity in scientific enquiry, and in gaining acceptance of his procedure as a standard approach he coupled rigorous maths with a whole series of practical tests and experiments on typical building materials.

Hooke's work – his experimental results cast into a simple mathematical model – form a conceptual bridge to the findings or inventions of two of the most influential mathematicians of modern times. Gottfried Leibnitz (1646–1716)[9] and Isaac Newton (1642–1727) drew together strands of thought centuries old to synthesise calculus, the mathematical modelling system that has remained the centrepiece of applied maths to the present day, only now yielding to other formulations because superannuated by its requirement for continuities of some kind in the processes it can be used to examine.

Vitriolic argument between Leibnitz, Newton and their followers over precedent for the invention of the calculus sharply divided subsequent developments between England and the Continent. By clinging to Newton's clunky formulation, English engineers trailed in the development of practical applications of this powerful tool for at least a century. Notions of an incompatibility between craft and theory lingered on.

The controversy was far-reaching. The Swiss family Bernoulli – Jacob (1654–1705), John (1667–1748) and Daniel (1700–1782) – fought Leibniz's corner brilliantly and in the process most rigidly defined how his thought would be transferred to use and what the common understanding of the method would be.

## Leonhard Euler

Other mathematicians of the time have proved only slightly less significant in directing the path structural engineering has taken. Leonhard Euler (1707–1783), a Swiss Calvinist from Geneva, was exceptional among exceptional mathematicians. His recognition of 'stationary points' within the continua of solutions available in the calculus enabled the prediction of the thresholds at which real world structures, columns, plates, shells and entire systems become unstable. This knowledge would be the key to designing 'lightweight' structures: geometry regaining its importance.

The pure elegance of the solutions and methods of solution to differential equations developed by Joseph Louis Lagrange (1736–1813) and Pierre Simon

Laplace (1749–1827) encouraged the belief that intrinsically beautiful forms could be generated directly by the automatic application of these algorithms. The real world asserted itself in the setting of boundary conditions necessary to solve the general equations, the designer exercising choice over the initial conditions from which the solution is generated. As mathematicians explored the calculus, more limitations came to light; it was shown that for some formulations no direct solution would be possible at all. This was vital to understanding the predictive possibilities of the maths.

Between these scientists and mathematicians an equation of conflicting sensibilities – Gothic, magic, Renaissance and Mannerism – was played out. The French encyclopaedist Denis Diderot (1713–1784) spoke of 'the sorcery of techne dispelled',[10] and this notion that art, science and technology were being uncoupled persists to this day, but no longer as a desired objective.

## The sublime

Besides abstraction, the natural world was being subjugated in two other ways. The refinement of designs for infinitely extendable estates and the sanitised bucolia of the English landscape garden directly challenged the profusion of the countryside. This was a direct register of a complicated notion surfacing at this time. Man's works transcended nature, not just for practical benefit but to create the sublime. In construction, stonemasons would study vaulting problems as perfectible geometric problems, completely divorced from gravity. Engineering was seen as an access point to this transcendental ordering, one that would describe a way without reference to physicality. The nature of perception itself was also being thrown into question. The French engraver Abraham Bosse (1602–1676), teaching perspective at the Académie Royale in Paris, declared that there was more truth in the axioms (of Euclid) than in our sensory perceptions.[11] Perspective and trompe l'oeil subverts our ability to actually see. As methods of depiction improved, with perspective projection mechanised, draughtsmen drawing more and more accurately and the first real world maps appearing, so a balance of the figurative, prescriptive and speculative gained in importance.

## Giovanni Piranesi and the Grand Tour

The engravings of Giovanni Battista Piranesi (1720–1778) displayed a new spatial sensibility, a step beyond Renaissance perspective.[12] They showed the deeper structure of things as well as their superficial appearance but without implying a veiled depth. This realism included meticulous studies of ancient construction methods, with the careful depiction of details, tools and techniques becoming the starting point for a sensuous exploration of classical construction. The artist didn't just want to catalogue Roman methods for contemporary use but wished to step into Vitruvius' very shoes.

Piranesi's *verduti* were part of a new way of seeing, a sensibility rapidly spreading across Europe. A scenographic separation was being established; there was the surface appearance of things, then the underlying way that that appearance came about – the two were separate. In architecture this became manifest in a separation of form from technical means, an attitude that subsequently has never really disappeared, and is flourishing again right now. Although these architects knew a great deal about construction, they separated themselves from it so that a second tier of artisans would set out and structure their buildings. A veneered space and set of elevations were supported by a 'black box' of structure beneath. As Newton described the mathematical limit of a function, so artists began to work towards the theoretical limit of their notions. Through linked series of theoretical studies and speculations, both Piranesi and Étienne-Louis Boullée (1728–1798) epitomise this phenomenon. Religion was waning in the face of a desire for understanding which focused on the immediate, the thing itself. The properly educated dilettante now brought back a set of prints from the 'Grand Tour', the perambulation around the sites of the ancient Mediterranean world necessary to complete a young man's education, but not just art and architecture were subjects of study. In his memoirs the Venetian adventurer Giacomo Casanova (1725–1798) shows off by mentioning that despite being in extremis he detoured on foot to inspect the dilapidated bridge at Narnia.[13]

## The Grubenmann bridges and John Soane

Not only classical ruins were sought out for examination. Particular examples of the new rationalism became well known and were visited along the route of the Tour, and students from northern Europe passing through the Alpine barrier stopped to see contemporary bridges and church roofs. Across the narrow, deep streams of Switzerland, crossings of middling length were required, which had to be financed by the small, localised communities. The bridge designers, most famously the Grubenmann brothers, Johannes (1707–1771) and Hans Ulrich (1709–1783),[14] would prepare display models as big as would just fit on a wagon and then hawk them along the riversides, stopping and calling for public subscriptions until a commission was obtained (**figure 6.1**).

The accurate miniaturisation of details in timber allowed for experimentation. Their origins lay in the roof structures of the simple Protestant churches of the Cantons in which the sequence of similar projects was exploited through measured but relentless experimentation, with a moral obligation to improve. The early examples are juxtapositions of simple frames, a principle that became extended into highly redundant trellises, cogged joints arrayed one behind another. A very modern form was tried, proto-laminated arches, but it proved to be ahead of its time and did not reappear until iron nails and bolting became commonplace a century later. These structures exercised a fascination for many architects of the time, such as John Soane (1753–1857), who prolonged his journey to seek out and sketch them. Their purity and artless response to a specific problem offered him a possible route to a new and rational building tectonics. He wanted to understand the underlying sensibility that gave them their artlessness as well as the particular carpentry methods they employed.

Soane seems to have been particularly responsive to the different attitudes to technology. Pursuing a close interest in vernacular building, he sketched

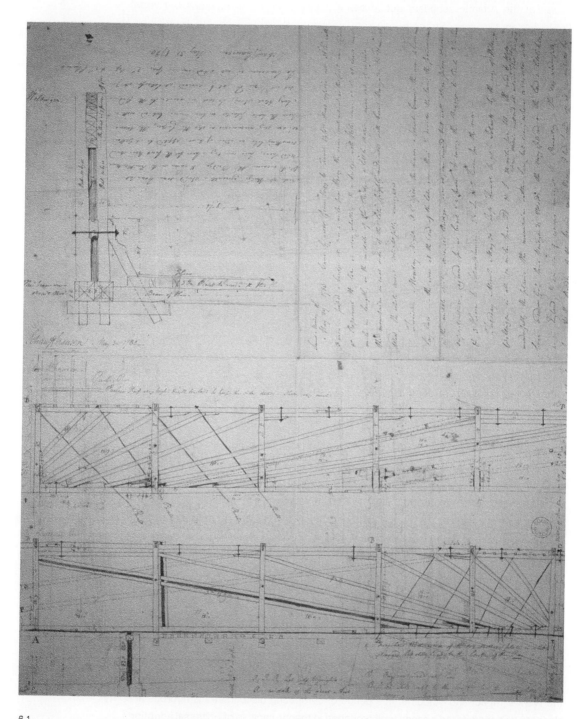

6.1

**Grubenmann Brothers, bridge design, 1758**

The Grubenmann brothers' entrepreneurial engineering allowed for a rapid concentration of development. Large-scale models were drawn by horse and cart to villages set alongside mountain streams. Purchasing the design – low capital cost, short lifespan design readily transformable into full-size components – offered trade advantage to impoverished burgers. The framing systems of consecutive bridges moved intuitively towards seemingly modern composite sections.

and collated examples then lectured at London's Royal Academy on 'truthful' construction, seeing in the way things are well made a way of obtaining an honest architecture. This sensibility persists today. His approach is best expressed in his work at the Bank of England, where he housed his clients and their celebrated mercantile attitude in a series of sober spaces reflecting their pragmatism and carrying it right through into the smallest of the construction details.

Just as travel expanded the minds of rich young Europeans, so the colonial encounter with new worlds forced other profound insights and appraisals. There was more time for speculation among those in virtual exile where the conceptual spaces offered in the strangeness of new environments released many individuals to think in new ways.

## Charles Coulomb

Charles Augustin de Coulomb (1736–1806) has been described as the first modern civil engineer. Born in the French provincial city of Angoulême, he trained in Paris, then the army and military academy, before being stationed on Martinique for nine years. This succession of thorough academic grounding followed by isolation in an underdeveloped environment produced in him a fully developed synthesis of applied mathematics and practical techniques. He made beautifully reasoned analyses of embankments, masonry piers and vaults. His method involved picturing potential failure mechanisms and was generated by close observation. The notion of an internal angle of friction is a homologue of what one sees in the field; the failure mechanism of a proposal had to be pictured, then the structure's resistance to it checked. In his diagrams the vestiges of geometric figures were present but were now transmuted into a graphical statics and crystal-clear model building onto which algebra and the calculus could be applied with relative ease.[15] With time on his hands, Coulomb was able to write refined, elegant and above all accessible expositions of his work. In this respect his example dispelled a lingering trait left over by the scholastics of veiling and obfuscating knowledge. Rather than write as a researcher only for other researchers, such men as Coulomb sought to clarify and expedite complexity for other practitioners to use safely. The engineering critic David Billington identifies this characteristic as belonging to only the best educationalists.[16]

## Colonial engineering

On his return to France, Coulomb made other contributions: in mechanics on sliding and friction forces, and in the latter stages of his career on electricity and magnetism. Many of his writings on mechanics were sidelined by political opponents only to be rediscovered after his death – an interesting process refracting and altering the work's impact and outcome. In a period of competing discourses his rigorous and clear expositions offered a canon well-grounded in theory, capable of universal adoption.

The colonial experience directed improvements in many technologies besides infrastructure engineering. Indigenous materials were treated more systematically and their idiosyncrasies exploited, and construction in tropical hardwoods called for the re-appraisal of established jointing techniques. The symbolic value of exotic materials was exploited. They were imported to demonstrate the wealth and abundance that the colonies produced and in turn the outposts needed consolidating as rapidly as possible using the authority of advanced technology. Sectional building and prefabrication techniques were resuscitated to meet the demands of the New World; western townships, churches and market halls were delivered in crates, to spring up almost overnight.

The vast distances being travelled made some age-old problems resurface. The sundial problem needing solution in so many latitudes provided an important impetus to descriptive geometry, parallel to but independent of perspective studies. Ocean navigation had been compromised by the difficulty of measuring longitude accurately. Reliable chronometers were required. John Harrison (1693–1776), the uneducated son of a carpenter, concentrated on clock-making and together with his brother James he revolutionised accuracy by making mechanical timepieces that did not require lubrication, uneven oiling being a chief source of inaccuracy.[17] Work on spring drives, unaffected by moving surroundings, eventually led him to create the first truly portable clock.

The measurement of time accomplished finally completed a static world. Harrison's refinement of mechanism, with its separation of power and control, began to shape the way modern machinery is ordered, drawing the support parts out to become an armature independent of the workings. Structure conceived as an independently functioning part of a larger system would have profound consequence for the role assigned to its engineers.

The designation 'Baroque' means distortion as opposed to symmetry, the effect that makes pearls globular rather than spherical.[18] Into the curves of over-stressed ideologies new methods were allowed to flow and condense. Shifts in emphasis allow the designation to be divided into early, high and late phases, but there was also a geographical distribution. Theocratic and political conflict arrived at different times in different places. It washed up in London as a cultural movement, but on the continent the Baroque's aspect as a bitter struggle predominated. European states were fully formed and wars had become truly international undertakings. Whole systems were in confrontation; Europeans and Ottomans faced one another. Politics became a shifting system of alliances, conditions in which it was worthwhile to construct major fortification systems around major cities. There was no point in resisting a siege unless an ally could be relied upon and allowed time to assemble an adequate relieving force. This aid would be remote, so road networks had to be good enough for the rapid movement of troops, but more importantly of the heavy ordnance which was escalating so much in scale. The armies of Baroque Europe fought in front of artillery.

## Artillery forts

Military engineering shaded into civil engineering. Conflicting requirements presented themselves; better communications, better barriers. The relentless confrontation between East and West ensured that the fortification of such 'frontline' cities as Vienna

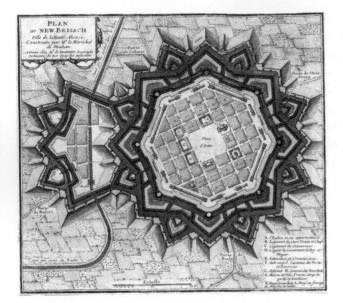

6.2
**Neuf-Brisach, Alsace-Lorraine, 1699–1709**
Imperialism underpinned itself with models of cosmic order. The encoding of attack and defence in salients and embrasures informs even the civic buildings. Vauban's methods were astonishingly effective; small garrisons could defend large perimeters, despite improving artillery. The success of the French south-east curtain encouraged tragic complacency later on.

remained cutting-edge technology.[19] Forts and earthworks were arranged to resist cannon fire; whole towns were now laid out under the weird geometries of the glacis and embrasure, fields of fire replacing anthropomorphic and pastoral space (**figure 6.2**). The value of engineers able to set out defences and the trench systems needed for attack was inestimable, and they became technical heroes. Sébastian Le Prestre, Marshal Vauban (1633–1707), adviser to the French king, Louis XIV, became one of the most admired military engineers of his age: 'Whatever he invested fell, whatever he defended, held.' He was widely emulated. His genius lay in spatial analysis, using models to assess topography and existing installations then modifying conditions to remove blind spots and indefensible areas. This three-dimensional form-finding called for a new appraisal of surveying, measuring, drawing and the manipulation of form.

## The Industrial Revolution

The turn of the eighteenth century in England saw the Industrial Revolution well under way. Wealth began to appear from the organisation of manufacture even before the development of the necessary power supply. Thomas Newcomen (1664–1727), an ironmonger supplying the Cornish mining industry, had a practical coal-fired steam pump engine going by 1712. Working by condensation generating an atmospheric pressure, the weak energy levels of the machine allowed large, simple components to be arranged in an appropriate form. Rising steam pressures then forced the improvement of structures to meet more highly stressed conditions. The first of the more efficient expansion-driven steam engines, unduly elaborate because its inventor James Watt (1736–1819) was too much the scientist, was patented in 1769. In the closing decades of the eighteenth century their widespread use began delivering seemingly endless quantities of power. The working through of the principles of thermodynamics, the theory describing their operation, followed on behind.[20]

6.3
**Spanish Town Ironbridge,
St Catherine, Jamaica, 1801**
The famous iron bridge in
Coalbrookedale presents an
exemplary homomorph, a timber arch
reproduced in cast iron. However, the
piece-part construction also suits the
inconstant material, and wedged
joints overcame primitive mould
inaccuracies. Prefabricated
construction allowed the structure's
reproduction in the West Indies,
contributing to the slave trade in
which early English manufacturing
innovation was so deeply implicated.

## Iron

Abraham Darby I (1678–1717), a Quaker metalworker, became obsessed with producing first brass, then iron, of consistent quality to use in mass production, goods for the common man. The members of his family were short-lived but together maintained a business with the continuity necessary to make wrought iron a practical building material. Darby's grandson Abraham Darby III (1750–1791) experimented with construction applications, directly substituting iron into forms developed for wooden or load-bearing masonry construction, not homomorphs but direct replacements, bridges and ships' hulls, which seem rather quaint now. These first steps of practical men needed the de-centring effect of the new learning to overcome their craft knowledge. Original form needed theory. It would be iron and its beautiful derivative, steel, that would exploit the potential of Hooke's discoveries. The conceptual tool had met its material.

The so-called 'iron masters' promoted their material with a very modern series of stunts, iron coffins and water closets. The Iron Bridge at Coalbrookdale, a masonry form with timber joint details all made of cast iron, is Darby's lasting monument (**figure 6.3**). Naturally enough, the design uses a repetition of load paths to overcome misgivings about the strength of individual components, but the first signs of a new filigree expression of structure is present in the array of attenuated rings.[21] This idealism, the demonstration of potential, is one of the strongest drives of the early days of industrialisation.

## Fireproof construction

The new manufacturing facilities, 'factories', were prone to fire. Cast iron beams in conjunction with brick vaulted floors were incombustible (**figure 6.4**), but early beams were brittle and downright dangerous and real perseverance was required to undertake the necessary improvement programmes. The development pressure from

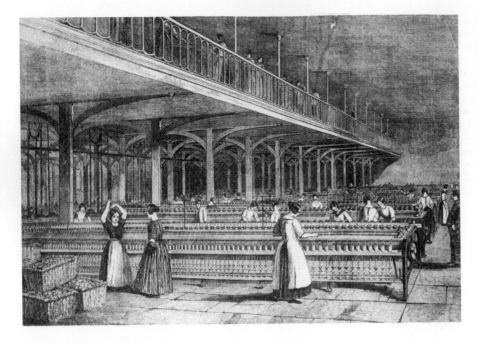

**6.4**
**Fireproof factory construction, 1810**
Early industrialisation consolidated around a new building type. Huge, repetitive mills were casings integral to ranks of looms and belt-driven machinery. The control of fire became paramount. Floors of brick arches, supported on but also shielding iron beams, were fire barriers as well as load-bearing surfaces.

the railways for reliable and ductile sections for rails and truck chassis filtered down to the construction industry.

In order to improve the new materials, testing machines were needed to describe and compare products. The unsegregated observation of behaviour was simplified into a series of standardised measures agreed by common acceptance. It became possible to specify quality rationally despite the nuances of real materials. Other properties disappeared from view as they ceased to be observable.

This rationalism had taken hold right across Europe. In its guise as the manipulation of information, it assimilated itself most firmly in France. The 'Lumière' taught that nothing was impossible. Bodies of thought were not immutable. Knowledge is protean and grows; the world would one day be described and fully mapped in an encyclopaedia. Ideas and intellectual techniques became saleable commodities.

## The Pantheon, Paris

A comparison of the Dôme des Invalides (commenced 1706) and the Pantheon (commenced 1758) in Paris (**figure 6.5**) shows how rational construction manifested itself in this milieu. Les Invalides is a neoclassical structure traditionally proportioned. Volumes are hollowed from the mass; a lightly framed lantern is carried on primitive timber trussing. The structure is unconcealed but within the mass. There is none of the inventiveness of Wren's brick cone system for St Paul's. In contrast, the design of the Pantheon seeks to fully exploit the technological means available to its time.[22] The interior space dissolves into a boundless forest of columns, a neutral void, and the main crossing piers are attenuated and disguised by pilasters. Structural logic is subsumed to architectural ideal so that there is a dislocation of structure and form.

Engineering devices, arches and oculi, are used to achieve concealed lighting effects. Structure is considered as a separate entity distinct within the whole, then its provisions are concealed behind the fabric.

The architect Jacques Germain Soufflot (1713–1780) studied the classical rules in the French Academy at Rome but kept up an interest in Gothic construction that was considered uncouth by his contemporaries. Practice in Lyon maintained his detachment. In his work he achieved his understanding by applying the latest structural analytical techniques to existing Gothic structures, and from that synthesising a new form of building. The Pantheon project was commissioned as a church but only completed in 1785, the year of the Revolution, and it became instead a mausoleum for the 'heroes of humanism', Voltaire, Zola, et al. The architect was

**6.5a and b**

**Pantheon, Paris, 1758–1789**

The first attempt at a completely rationalised structural design, a church design switched to a pagan hall of heroes during the French revolution; every part is sized by calculation. The main piers spalled badly. The assumption that full areas of stone carried load was invalidated by the masons' practice of cutting only corners precisely.

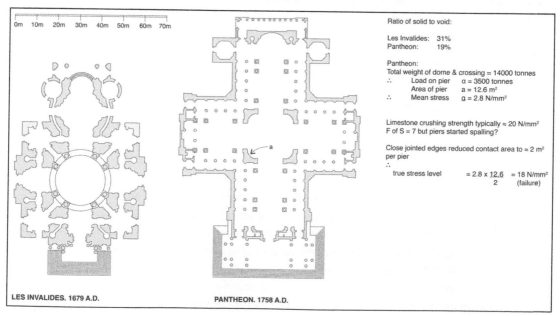

0m 10m 20m 30m 40m 50m 60m 70m

Ratio of solid to void:

Les Invalides: 31%
Pantheon: 19%

Pantheon:
Total weight of dome & crossing = 14000 tonnes
∴ Load on pier $\alpha$ = 3500 tonnes
Area of pier $a$ = 12.6 m$^2$
∴ Mean stress $\sigma$ = 2.8 N/mm$^2$

Limestone crushing strength typically ≈ 20 N/mm$^2$
F of S = 7 but piers started spalling?

Close jointed edges reduced contact area to ≈ 2 m$^2$ per pier
∴

true stress level $= 2.8 \times \dfrac{12.6}{2}$ = 18 N/mm$^2$ (failure)

LES INVALIDES. 1679 A.D.

PANTHEON. 1758 A.D.

four years dead by then. Despite his reasoned approach, the building's completion was dogged by the progressive development of structural defects in the piers. The foundations were well thought out, and remote buttressing was deployed appropriately. The problems seem to have stemmed partly from bad construction technique accumulated within the Parisian contracting system and partly from over-innovative detailing.[23] What appeared to be better and more logical ways of assembly fell foul of the unforeseen. A new way of dressing the load-bearing masonry resulted in serious spalling and a weaker construction than had originally been allowed for.

As is so often the case in remedial work, there was time and conceptual space in the situation to consider the engineering at length and speculate on new ways of addressing the problems. The large dome was supported on the slenderest group of columns with an excess of ironwork to bind and contain the base rings. Emiland-Marie Gauthey (1732–1807), a celebrated teacher at the École des Ponts et Chaussées and who completed the Canal du Midi, used graphical statical methods to prove that Soufflot's original section sizes were adequate as they were so that only the joints needed amending. Relieving structures were added nevertheless, designed by Jean-Baptiste Rondelet (1743–1829) of Lyon, another champion of rational construction, but one who took a direct route to securing Soufflot's ambition.

## Building renovation and new forms

The study of old structures for precedent shaded into an interest in the renovation of monuments.[24] Many of the old cathedral establishments were now subjected to their first major refurbishments, and a re-emphasis on heritage is reflected in widespread repair works. Visible elements were restored in their original form, but other structural systems were upgraded to contemporary practice and any inherent faults corrected. The changes made to timber cover roofs particularly illustrate the advances in structural understanding that had been made over medieval times. Trestles and partially braced frames were modified logically, and fully triangulated trusses with members aligning at the nodes, so-called 'axially-loaded' structures, became the norm, the most perfect and efficient way of assembling small components.

Inexorable industrialisation and the growth of mercantilism promoted new kinds of building. The merchants' houses in such cities as Amsterdam were sober and understated, responding directly to the environment with large windows and tall comfortable rooms. The simple technologies of brickwork and stone were honed and refined; very large halls, cloth markets and assembly rooms were structured without drama, relying on clarity of form. Proportion was a combination of classical and practical dimensioning, and hierarchical organisations of supporting elements were often concealed behind neutral envelopes with everything cleverly fitted in. Structurally these arrangements presaged the major–minor frame systems of modern buildings. The wrapped façades, colonnades and cloisters are pre-adaptations of the modern curtain wall, distinct from its backing.

The American colonies were burgeoning. The abundance of timber and a climate benign to its use (the relative humidity below 20 per cent discourages rot), coupled with a freedom of expression and a new professionalism emphasising clarity and rationality of thought, created design difference.[25] Simplicity was often very studied so that seemingly naïve joints and details might conceal excesses of craft. New tree species and their environments provided very large timber and structures tended to be bigger and more powerful than European precedents, particularly the ships. The on-going shortage of manpower meant that expedients to reduce labour quickly became a preoccupation. The simplest timber framing carrying clapboard linings, all produced rapidly from the most rudimentary lumber mill, became the staple of the new country. Special buildings for the internal colonisation were ordered from a catalogue (**figure 6.6**).

So the stage seemed set for great advances both in North America and in a Europe where the full establishment of rationalism and its application of technology were directly consequent to the French Revolution. And so it was to occur at first. For a blessed period of time the underlying assumption that engineers relied upon was an unadulterated positivism, holding that the only authentic knowledge is scientific knowledge. The central principle was probably best formulated by Laplace who said:

> We may regard the present state of the universe as the effect of its past and the cause of its future. An intellect which at a certain moment would know all forces that set nature in motion, and all positions of all items of which nature is composed, if this intellect were also vast enough to submit these data to analysis, it would embrace in a single formula the movements of the greatest bodies of the universe and those of the tiniest atom; for such an intellect nothing would be uncertain and the future just like the past would be present before its eyes.

6.6
**Corrugated iron Mission church, 1858**
Colonial expansion consolidated itself with institutions. Public buildings, churches, hospitals, schools and courthouses, mass-produced by industrial methods, appeared almost overnight, suppressing vernaculars and proclaiming the new order. Skills could be withheld from the local populations.

AN OLD CORNISH RAILWAY VIADUCT,
TRENANCE NEAR ST. AUSTELL

Chapter 7

# Encyclopaedia (1750–1860)

## Napoleon

Napoleon Bonaparte (1769–1821) wanted to improve the whole world and sought to impose the secular ideals of revolutionary France throughout Europe and further afield. His tactics involved sacrificing huge numbers of his allies,[1] crushing victories enabling him to implement a programme of total control over those who opposed him. The Corsican artillery officer could only follow his military cast of mind by codifying all things to bring civilisation into a single mould. The catalogue of conflicts shows how the old monarchies resisted but were infiltrated by subtler processes nevertheless. Ideas of all-encompassing order, social justice and man-made salvation were discussed and assimilated in England, Russia and the German lands.

The universality of the Napoleonic programme and its 'scientific' grounding were fundamental to it. Everything was to be reconsidered and regularised to an optimum performance, irrespective of traditions or intangible benefits. Tree-planting initiated throughout France in the First Empire was all with plane, an alien species from the eastern Mediterranean, of exceptional longevity with no European parasites and a shedding bark resistant to urban pollution. This technical idealism was occasionally debased by propaganda, a subtext promoting change for its own sake or as a demonstration of power. Except in Britain and Scandinavia, the European custom of passing on the right (sword) hand was changed now to Napoleon's left.

The thoroughness with which this outlook was promoted is reflected in Napoleon's Egyptian expedition of 1798. The invasion was part of an ambitious military strategy to strike at Britain through its eastern connections. His army was attended by a complement of technologists and artists whose investigations were to shape Egyptian archaeology and also a pattern of research and analysis into which this book itself fits.[2] The physical products of the past were to be read as a coding of other consciousnesses, with records, relics, beautiful engravings and watercolours all arranged to reconstruct a lost cultural system. Artefacts as well as literature and social structures were meaningful.

7.0 (opposite)
**Postcard of Trenance Viaduct, Cornwall, 1859**
An undercapitalised rail company racing to open its route promoted a unique set of structures crossing the wooded valleys of Cornwall. Frames of Baltic pine were to be replaced with more permanent brick arches once the money flowed. A variety of designs show Isambard Brunel exploiting the opportunity to experiment.

## Meritocratic engineering

A new generation of technologists needed training: new techniques, new attitudes. The egalitarianism of the Revolution encouraged novel engineering schools and initiated a genuine meritocracy. Gaspard Monge (1746–1818), son of a pedlar, had been refused an army commission because of his low birth but quickly rose in status as the Revolution progressed, escaping the Terror to become a university professor. He was with Napoleon on the Nile, and stayed close to the Emperor even after Waterloo. More than those of any other individual, his initiatives have shaped the condition of today's general practice and our current education systems are structured around the curriculum he developed for the École Polytechnique, the school of civil engineering he was asked to establish by the National Convention.

## Engineering education

The organisation of the new education system acknowledged the precedence of military over civil engineering. Various departments were separated out but then recomposed in a uniform way, relying on consistent methodologies. Practising engineers and theoreticians were recruited in equal part to teach in and run the new schools; important contributors included Gauthey (see page 102), Culman (see page 131), Jean-Victor Poncelet (1788–1867) and Claude-Louis Navier (1785–1836), who all wrote teaching manuals based on their researches. The modern textbook invented itself as coursework consolidated. A standard treatment proceeding from general principles towards specific applications in an exposition supported by worked examples became the norm as these men attempted to systematise European engineering. Absolute excellence and elitism became established components of European engineering.

Just as important as the creation of model teaching institutions was Monge's development of technical drawing. Snubbed by the officer's school, he sought advancement as a mathematician, teaching himself while employed as a draughtsman. He saw a simpler way of setting out fortifications than the contemporary methods. As his academic star rose, he realised that he could systematise a way of drawing that could describe and communicate anything; objects could be fully recorded or perfectly reproduced. The rules of orthogonal projection set out in his treatise *Descriptive Geometry* (1795) are still our standard:[3] three drawing planes at right angles to one another on which to project a point are sufficient to hold it exactly (**figure 7.1**). Although any shape of object can be created or recorded, Cartesian coordinates prefer rectangles, and the characteristics of this descriptive tool have trammelled much modern design.

## Claude-Louis Navier

The career of Claude-Louis Navier well illustrates the composite activity that engineering had become in revolutionary France. He was born into a bourgeois family in Burgundy and his education was closely supervised by Emiland-Marie Gauthey, his uncle. Emphasis was placed on taking theoretical studies in physics and making them

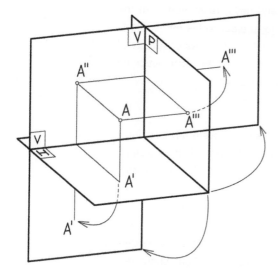
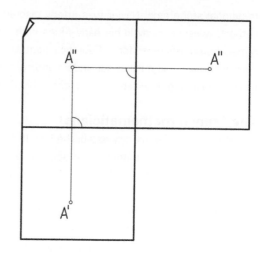

7.1
**Gaspard Monge's descriptive geometry, 1798**
Taking the complexity of a three-dimensional object, projecting and recording it onto three orthogonal planes, unleashed not only a universal means of communicating design complexity, encoding for machine manufacture, but also a new way of seeing, constricting as well as enabling.

directly applicable to practical problems. Such studies were disseminated by the publication of treatises. Though highly learned, these manuals reveal a progressive elimination of errors and false assumptions aimed towards establishing an all-encompassing method. There is an absence of bookishness; first principles rather than precedence or research guide their findings.

Navier seems to have been unaware of and therefore trailing the sophistication of Coulomb's work of fifty years earlier. He secured a commission to build an exemplary suspension bridge across the Seine in Paris. Form and proportions were to be ideals determined by calculation; catenaries – the curves taken up by hanging chains – yielded elegant transcendental functions upon which to ponder. The ensuing debacle stemmed from two sources, practical and political.

Construction commenced on a design whose elements had been refined towards their theoretical limit.[4] In these regions there is little to spare on the unforeseen. Navier displayed a propensity to solve problems amenable to clever maths while ignoring more immediate but prosaic concerns. He solved the equations governing longitudinal vibrations in the main suspension bars, an irrelevant problem, but was unable to overcome the differential heating problem of one chain shadowing another. This secondary effect caused serious imbalances in the hanger assemblies. During the erection of the first chain, one of the abutments moved. The small con-solidation could have been accommodated but confidence waned, and here the second source of his downfall prevailed. His 'new designs' had been met by a con-certed conservative reaction. The work was generally sound and the teething troubles of innovation all surmountable; his insistence on minimal design, however, left him little manoeuvring room. He was not allowed the time to resolve the technical situation into which he had projected himself. Instead of adjustment and amendment, his critics forced on him dismantlement and the replacement of his vision with a traditional stone arch. Navier was utterly discredited and never recovered from the setback, wasting away in his early fifties with no major project to his name. He is remembered, however, for a theoretical contribution; the Navier-Stokes equations of fluid dynamics, essential to our understanding of flight. He found that formulation by flawed reasoning: intuition in the midst of his rigorous process.[5]

## The French mathematicians

A mathematical meritocracy established itself in France, a social system promoting a host of individual contributions against a common conceptual background. Constraints were just sufficient for the work to be synthesised into all-encompassing continuity while not preventing extraordinary diversity. A strain of uninhibited amateurism projected mathematicians beyond the bounds of mainstream academic work and careers were being made by convoluted ascents through the academies. A cursory listing demonstrates the steady condensation of ideas that is the modern theory of elasticity, universally applicable to structural problems. Four characters stand out. Their contributions to modern engineering are bound by a common thread of idiosyncrasy.

Augustin Cauchy (1789–1857), brought up in Paris through the worst of the Revolution, was a difficult, obsessively religious man, arguably one of the greatest mathematicians ever. He spent much time arguing with and competing for promotion in French institutions,[6] where his wide-ranging studies enabled him to make connections between disparate enquiries. He founded complex analysis, the most powerful branch of applied maths, and initiated the mathematical theory of elasticity in its contemporary form. A modulus linking stress and strain had been proposed by Thomas Young (1773–1829), an English Quaker and polymath, which, combined with the notion of strain compatibility, became the basis of Cauchy's mathematical treatment.

Barre de Saint Venant (1797–1886) was another mathematical prodigy. As a conscientious objector during the revolutionary wars of 1814, he was ostracised and worked feverishly for recognition. He wanted results that would be widely applied. Confining himself to the theory of elasticity, he obtained many of the earliest and most important exact solutions for the analysis of iron structures.

Meanwhile, Denis Poisson (1781–1840) escaped poverty through the new education system to become a teacher of physics and mechanics. The ratio named after him enables the three-dimensional effects of strains and vibrations to be considered. Noticing the correspondence between the equations describing heat flows through solid bodies and strain fields, he initiated the idea that stresses can be imagined as flows of force.

Jean-Marie-Constant Duhamel (1797–1872) rounded off the basic theory by defining its limitations. Mathematical models of real situations could be constructed, but not necessarily solved. Duhamel identified tests for these insoluble forms, which would have to be tackled by other means, the iterative procedures and successive approximations that are suited to mechanical calculators.

## Stability

Along with a better understanding of strength and deflection in structures came corresponding advances in the assessment of their overall stability. The subject had comprised a hotchpotch of ways to assess how and when structures buckle, suddenly shedding their strain energy. Some were based on the calculus, some on graphical statics, and others on a combination of the two. The Swiss mathematician Leonhard Euler (1707–1783) showed that the singularities exhibited in some differential equations could represent buckling behaviour and derived the critical loads at which struts would suddenly bow. Lagrange and Gabriel Lamé (1795–1870) extended this approach to complete frameworks, trusses and domes, Lamé in particular making practical use of his French education during a sojourn in Russia. His systematic studies of the locally produced iron resulted in a detailed framework for describing the properties of elastic materials. However, account had to be taken of real world imperfections and the residual stresses of manufacture if the theories were to be of real use.

Notions of universally applicable principles came to building engineering from other sources besides mathematics. The French philosopher and polymath Denis Diderot (1713–1784) had set out to produce an encyclopaedia containing all knowledge systematically arranged. The construction counterpart was *Traite theorique et pratique de l'art de batir*, in which Rondelet (see page 102) divided built form into a set of irreducible forms, capable of infinite combination.[7] These could be structural stereotypes, domes and vaults, as well as circulation patterns, colonnades and axes. This approach maintains a presence today, resurfacing in the taxonomies of Christopher Alexander's *Pattern Language*[8] and Francis Ching's *The Elements of Architecture*.

## Structural metonymy

An important assumption of these structural taxonomies originated from the scientist Georges Cuvier (1769–1832), who recognised the record encoded in fossil remains and geological strata. From this he inferred the existence and extinction of types and the metonymy of part and whole.[9] A knucklebone was sufficient to conjure up the appearance of a whole dinosaur; archaeologists used fragments of fluting to infer the layout of entire temples. Within this work the concept of morphology took shape. The form of living creatures reflected their environment;[10] the shapes and details of buildings encode their surroundings and the preoccupations of their makers.

## Eugene Viollet-le-Duc

As well as determining new ways to build, French architects addressed the existing built fabric. An extraordinary heritage of medieval and ancient building surrounded them, but how was it to be appropriated and renovated for the new age? Eugène Viollet-le-Duc (1814–1879) sought to identify a rational continuity in historic building, to get behind the styles to the technical means that generated them. This he did by re-investigating medieval Gothic construction purely as a paradigm of rational tectonics; new materials, iron and glass, would be used within the grammar of Gothic to create a new architecture.[11] Disseminating these ideas in written polemic and paper projects, he secured a number of over-imaginative restoration projects and quirky new buildings. These were powerful mental tools being forged by high theory, and they needed to be. In Britain the products of the Industrial Revolution were being relentlessly redeployed in ever accelerating expansion and development. A group of practical, hard-nosed engineers cum businessmen appeared to harness theory to practical problems of ever increasing scale.

## English (and Scottish) pragmatists

Thomas Telford (1757–1834) was a balanced and pragmatic engineer, a great road and canal builder, persistently improving on bridge systems over a long and busy career.[12] Despite projecting himself as conservative and reliable, he could on occasion propose projects of almost fantastic scale. His practical style, blending proven means with the simplest theoretical methods that would serve, became very influential in Britain, where he has been styled 'the father of modern civil engineering'. He was not above besmirching rival engineers' reputations with dubious objections and whispering campaigns.

Several of the trunk road alignments crossing Britain are Telford's almost unrecognised masterpieces of civil engineering, but it is the high chain suspension bridge taking the London–Holyhead post road across the Menai Strait for which he is remembered (**figure 7.2**). Relatively cumbersome in comparison with contemporary French examples, the crossing relies on the weight of the heavy link plates for its stiffness. On completion, the light timber deck would flutter in high winds and the roadway was progressively stiffened with side trusses. Better understanding of relative stiffness between deck and suspension system was won by close observation in service and protracted amendment and adjustment.

Another suspension bridge pioneer, Captain Samuel Brown (1774–1851), came to their design from another background. After Waterloo, the overmanned British navy needed outlets, polar exploration or Far Eastern trade. Brown brought an expertise in ships' rigging to a structural problem not yet fully circumscribed with received knowledge. His technology transfer created attenuated suspension bridges of astonishing grace and sometime unsound delicacy, quite unlike Telford's stolid response. Under dynamic loading, his footbridges and railway bridges might display excessive deflections; on one bridge the locomotive would push the deck like a wave before it. The behaviour of suspension systems under partial loadings was made evident only so to be manipulated and controlled.

## The Britannia bridge, Menai Strait

Robert Stephenson (1803–1859) was the most important of the railway pioneers. As well as steam traction and the refinement of the permanent way, the use of hydraulic power, control systems, large bridge structures and other civil engineering works were all transformed by his prolific and inspired output. In a well-judged collaboration, he

**7.2**
**Menai suspension bridge, Anglesey, 1826**
Thomas Telford's masterpiece completed the London–Holyhead road, a military highway, while allowing the Admiralty sufficient air draft for its tall ships. The primitive catenary chains, overburdened with links and pins, are sufficiently heavy to stiffen the system.

produced arguably one of the first truly modern structural designs. The line of the London to Holyhead railway had to cross the Conway Firth and Menai Strait. Around Anglesey the Admiralty, lobbied by Liverpool shipping interests, imposed a stringent airdraught (clearance beneath bridge) precluding most conventional structures. In response, Stephenson came up with a completely new structural form, the tubular bridge:[13] the train would run within a hollow reed of wrought iron plate (**figure 7.3**). He tested his proposal, then set out its theoretical criteria, identified the potential mechanisms of failure and finally exploited the latest technologies to efficiently manufacture his design. He worked through and determined the exact material and manufacturing quality requirements for the fabrication. In these processes he sought the help of experts.

William Fairbairn (1789–1874) had been testing full-size metal girders and fabrications, calibrating practical observations with theory and simple rules of thumb. He was enlisted for a series of large-scale model tests. For analytical justification of his sizes, Stephenson went to the farmer turned mathematician Eaton Hodgkinson (1789–1861), well versed in the latest theoretical developments being made on the Continent.

7.3
**Britannia bridge, Menai Strait,1826, by Robert Stephenson (1803–1859)**
The utter pragmatism of the design coupled with the apparent flexibility of the engineer's approach seems to achieve a thoroughly modern aesthetic. The box girders proved so stiff that the intended chain supports were simply left out. Even the Egyptian decoration has an elemental plasticity.

In fact tubular bridges required a profligate use of materials and were a stop-gap response, co-existing with early yet far more efficient forms of space structures.

Continuity was a key structural concept of the Menai design, and the multiple spans were connected to form continuous girders, thereby reducing deflections and weight. Erection sequence and pre-cambers were carefully worked through to ensure that the moments over support and at midspan were balanced in magnitude. British engineers had previously been too contemptuous of theory, and this prediction and manipulation of forces at construction stage presented a new outlook. The large number of riveted plate elements required in the design were manufactured using an automated process. This was controlled by a punch-card system adapted by Richard Roberts (1789–1864), from Joseph Marie Jacquard's (1752–1834) invention of 1801 for guiding power looms.[14] That precocious use of automatic control provoked the original Luddites.

The construction of the single-span low-level Conway bridge provided a good test bed for the much more difficult Menai crossing. Stephenson's erection procedure was a tour de force. The main beams were constructed complete on the shore then, having floated the tubes out beneath the supporting piers, he used steam-powered hydraulic rams, his favourite technology, to lift the prefabricated girder assemblies in short increments, bricking up behind until they reached their final level. The structure was stable throughout this sequence, constructed in a perfectly fail-safe manner.

The scale of the Menai project influenced contractual procedures and required many innovations in the organisation of the workforce. The management of skilled labour on massive yet relatively short-term undertakings had to temper exploitation with rudimentary concerns for health and safety if the nation's procurement base was to be maintained. The adversarial contractual arrangements of the period persist to this day, but the uncertainties of the new scale of construction initiated attempts at more rational risk apportionment.

## Contingent design and immediate obsolescence

Despite the long-term serviceability of the Conway and Menai bridges, hollow girder configurations were short-lived as design solutions, made redundant by alternatives far less profligate in material and construction resources. Tubes were rapidly relegated to component parts of ever larger structures, such as Brunel's Tamar crossing[15] and later Fowler and Baker's Forth railway bridge (**figure 7.4**). The success of an engineering solution can be assessed in many ways. Stephenson's achievement is undeniable; in technical terms his design has been styled 'a desperate remedy to a desperate situation'. The form proved to be an evolutionary dead end, however, as bridge design moved on in other directions.

## Isambard Brunel

Stephenson's contemporary Isambard Kingdom Brunel (1806–1859) is established as the epitome of the Victorian engineer.[16] The man in the stovepipe hat continually chewing a cigar as he surveys his work is a stereotype set alongside the pith-helmeted explorer and deer-hunter detective in defining an age. Brunel worked in a competitive commercial environment and indeed recognised the value of flamboyance and self-promotion. He produced an extraordinary range of beautiful, joyful and extreme structures, ships and devices.

His father, Marc Isambard Brunel (1769–1849), had had a celebrated career developing civil engineering methods and mass-production processes. The son was sent to France to obtain a theoretical grounding before returning to the family business. Once the collaboration had established itself, Marc retired to become consultant. The projects undertaken by the two men are characterised by an extraordinary confidence, almost bravado. The Thames tunnel, fraught with setbacks and fatal accidents, was on the limit of practicality, inserted through a thin stratum immediately beneath the river bed.

Isambard's early competition win for a suspension bridge over the Clifton gorge was completed posthumously. It was not the first prize winner; that scheme disappeared mysteriously. A very modern design sensibility is revealed in Brunel's proposal, a stripped-down handling, whereby nothing detracts from the majesty of the engineering.

There are also occasional notes of technological excess introduced into his projects to draw attention to themselves. The level of the Great Western railway on its run out of London needed to stay as low as possible for economy and this is made manifest where the line crosses the Thames at Goring. The crowns of the elliptical brick arches of Brunel's bridge are almost un-reasonably thin and he relished both the controversy aroused by the design and the subsequent humiliation of his nay-sayers when vindicated by proving engineers.

Brunel's propensity to project himself into extreme situations inevitably courted disaster. His broad gauge was marginalised, the atmospheric railway abandoned and his underpowered locomotive designs were replaced. His most famous reversal was the *Great Eastern*: the leviathan steam ship intended to ply between England and India was simply a precocious folly. The energy balance just did not add up for its projected mission down to India along a coal-less route.

## Engineering showmanship

Brunel was neither generous nor fair with his collaborators, and relentlessly exploited PR opportunities. His portraits in front of the vast drag chains at the launch of the Great Eastern are one of the earliest uses of the photographic medium for promotional purpose, the variety of poses tried out indicating the intention behind the process.

In collaborative work Brunel kept his consultants subservient. Unsound in naval architecture, he gave his building projects an elemental directness by this practice. Waterloo station, London, is a clearly resolved pattern of structure: clever, simple detailing such as hardwood levelling wedges to the wrought iron arches, all embellished with a superficial tracery by the Gothicist Matthew Digby-Wyatt (1820–1877). The architect's introduction of a transept bisecting the train shed leavens an otherwise leaden set of arcades.

## The Crystal Palace

Two years after the opening of Waterloo station in 1848, Prince Albert's Great Exhibition was staged in Hyde Park. Brunel was on the steering committee. The temporary building was designed and completed in nine months by the gardener Joseph Paxton (**figure 7.5**). The prototype was a glass enclosure housing a giant Amazonian water lily. The pattern of the plant's ribbed leaf was supposedly transposed onto a simple ridged glass roof then rolled out to make a massive diaphanous enclosure. The absolute

7.5

**The Great Exhibition building, Hyde Park, 1851**
A glazed shed containing the fruits of empire and industrialisation, the translucent, transitory and weightless envelope has been cited as a new architectural sensibility. Designed by the gardener Joseph Paxton for rapid construction and maintenance accessibility, inadequate lateral stability of the attenuated trabeated frame made the original building dangerous. The most interesting feature to modern eyes might be the junction of infinitely extendable grid and special barrel vault retaining two trees protected by public outcry.

austerity of the competition-winning scheme was debased by an arched centre added to save some existing trees. The undecorated and mechanistic detailing, careful account of construction procedure and demountability made the building the quotable touchstone of architects interested in high technology during the 1970s and 1980s.

In 1855 Brunel designed a sectional hospital building for the British army in the Dardanelles, a tour de force of engineering integration.[17] Secured through sinecure, the project is made into an exemplar of integrated engineering, environment provisions, ventilation and heating, structural panels and roof trusses all consolidated in a unified whole.

## Steel

The Crimean War was the occasion for a critical leap in materials science. The call for ever heavier weapons and the armour plate to resist them led Henry Bessemer (1813–1898) to invent his eponymous process for the production of cheap, high-grade steel. Boilers could be made to much higher strengths and so run at much higher pressures. Efficiencies tipped over to the point where worldwide cargo-carrying under steam power became practical and economic. This benign consequence of the conflict dispersed a material that behaved consistently and reliably according to the ever more complicated calculations used for design predictions.

## The Tay bridge disaster

The increasing accuracy to which structural behaviour was being assessed called for a corresponding refinement of loading estimation. Only if the environment surrounding an artefact were properly understood could the benefits of better method be realised. The disaster of the Tay bridge collapse (1879) demonstrated how the assessment of loading patterns, particularly wind pressures, had lagged behind stress analysis when, two years after the bridge's opening, a train and all seventy or so passengers were lost in its collapse. There was the ever present human frailty of bad workmanship – many of the cast iron components were found to have serious flaws concealed with 'Beaumont's egg', a precursor of Polyfilla – but the engineer, Thomas Bouch (1822–1880), had also underestimated the north wind from Scandinavia.[18] His general arrangement and details lacked robustness. The designer did not recover from his indictment, and he died shortly thereafter.

Statistical tools, developing for scientific investigation and social manipulation, proved invaluable to deal with the uncertainties of weather and events. Applied to physical testing, the same techniques allowed the experimental strength of materials to match theoretical studies with practical values.

## Cultural differences in engineering

In North America commercial conditions drove structural engineering development differently than in Europe, and suspension bridge design, for example, progressed along the most pragmatic of lines. The American patent system made a protected system of professional consultancy unnecessary. Instead of engineers scratching for fees money was to be made by developing schemes that could be repeated or 'rolled out' without continual specialist input. James Finlay (1762–1828), born in Maryland of Scottish stock, experimented and perfected a proper deck-stiffened suspension bridge system. He standardised anchorage details and hanger junctions before down-specifying materials and workmanship to make it safe to license others to construct his bridges across a whole continent.[19]

A cultural divide in design and research between Europe and the New World was being established. North American academies were modelled on European prototypes and engineering schools followed French precedent. Though education maintained a common thread, professional status was measured differently on the two sides of the Atlantic. The formalisation of engineering training in Britain, France and Germany took on varied emphases. Research and theory predominated on the Continent, while Britain encouraged her students to make earlier links between practice and theory.

## Extreme clippers

The migration of ideas and technologies across the American west dispersed into a banding of uptakes dependent on local conditions, a kind of chromatography of reception. European concepts were ransacked for usefulness, dislocated parts from which to make something indigenous. These processes were partially short-circuited by the technical developments made to enable the sea route to the West Coast, round Cape Horn. The Californian gold rush of 1849 stimulated the world's economy, heralding a decade-long boom and revolutions in transportation, mercantilism and cultural mixing, a destabilisation culminating in the American Civil War. The geography of the mass hysteria promoted a beautiful niche technology, and Yankee clippers – fast cargo vessels – delivered goods and information directly from the Eastern seaboard to the south-west coast. The design of these ships far exceeded practicality: space sacrificed for speed.[20] Extreme clippers such as Donald Mackay's (1810–1880) *Flying Cloud* were driven almost to destruction, regularly losing gear, just to achieve passages between New York and San Francisco of less than ninety days.

This technological aestheticism spread both sides of the Atlantic. The British raced for tea, cotton, jute and finally grain from Australia; the Germans brought a sobering influence to sailing ship design, seeking nitrates from South America.[21] As a transport system soon to be superseded by steamboats, large deep-water sailing ships completed their development potential during an extended period of decline. Iron and subsequently steel construction would be shipbuilding's future. Naturally enough, naval architects offering new materials initially made direct substitution of metal for wood over existing forms. Hull structures were composites; the *Cutty Sark*, an extreme tea clipper of 1869, has a timber hull sheathing over metal frames.[22] Steamship architects were less constrained by tradition, and directed the development of wrought, rolled and cast products to their own ends, integrating heavy engines and hull structures for the new forms of powered ship. Sailboat yards then in turn appropriated plate work, gradually moving towards new scales of vessel.

Steam winches allowed more canvas to be handled by fewer men, prolonging the survival of this labour-intensive industry.[23] The writer Joseph Conrad (1857–1924) uses the environment of an iron sailing ship as the site for an examination of modernism replacing an older sensibility (**figure 7.6**).

Shipbuilders were handling the larger magnitude of forces generated by the new engineering as the size of their particular work doubled and doubled again. The shipwright was now able to set out an extremely hydrodynamic hull and deal with the mechanical power required to move vessels upward of some 500 tonnes. The insertion of marine steam engines, then the introduction of propellers, meant that ship hulls needed a new level of strength to withstand being driven through high seas, forced against wind and current across the North Atlantic and around the Southern Capes. The process of reducing weight and raising structural efficiency of the parts was assisted by low reliabilities, making continual observation and replacement necessary. Resources were carried on board to continually replace the ephemera of the upper works.

## New load environments

The highly stressed conditions met by components of transport systems called forth a new set of structural problems. The use of strong alloys in rotating machinery or heavily loaded moving parts such as railway wagon axles and rails revealed the phenomenon of fatigue: failure below the anticipated material strength as a result of

7.6
*Narcissus, 1876*
Built on the Clyde in 1876, the 1,270-tonne vessel is painted with the chequerboard of an East Indiaman, the vestige of a defensive gesture, appearing like an armed ship to ward off pirates. Seamen detested the dangerous over-sparred vessels, their benign barques suddenly industrialised.

load fluctuations. The first studies of this problem used experimental data to predict breakage points, then mathematical physics was called upon to explain the processes involved. Metallurgy was informed by these studies and material properties manipulated in a resurfacing of alchemic sensibilities. This work would be critical to aviation design, only eventually filtering down to general structures, particularly bridges, as material properties and consistencies improved.

# Chapter 8

# The American reconstruction (1860–1890)

The American Civil War (1861–1865) has been represented as a very special stage in humanity's development, the first fully industrialised conflict.[1] Victory depended on economic strength, with superior technology critical to the struggle, and mundane improvement and outlandish innovation proceeded together. It seems as if almost every possibility for advantage was grasped at. Gunsmiths Horace Smith (1808–1893) and Daniel Wesson (1825–1906) patented the metal cartridge in 1852, ensuring the power and accuracy of firearms and paving the way to automatic weapons. Huge conscript armies required cheap, simple and reliable rifles. The Union army relied on the Springfield Armoury set up by the industrial designer Eli Whitney (1765–1825),[2] and it was there that the essential steps towards fully mechanised mass production were taken, bringing manufacturing tolerances to a level where interlocking parts could be made interchangeable.[3] Meanwhile, especially in the weaker Confederacy, curious marginal and hybrid designs, ironclads and submarines, were tried out in desperation, pre-adaptations of later forms.

Post-war reconstruction, slowed to the pace of the economic recovery, allowed a more considered selection and concentration on systems development. The resurgence in building called forth a new type of engineer and a new scale of engineering as a nation of survivors – almost indifferent in their confidence, grimmer in motivation and utterly unawed by the new means – seized their opportunity.

## Commercial engineering and the Roeblings: father and son

The two generations following the war established a distinctive North American way of engineering. It was a time of pragmatists. The expanding economy needed technical solutions of breadth, capable of wide-scale application. The Roeblings, father and son, epitomise the post-bellum engineers, staring out of their silver nitrate portraits, proud and self-possessed in their dusty, crumpled clothes.

8.0 (opposite)
**Brooklyn bridge, 1863**
A group portrait is taken on the north anchorage during construction. Cable-end connections are the critical details in all suspension bridges. The ability of these men to find and confront only key issues underwrote their achievements.

John Roebling (1806–1869) started out as a mining engineer in Silesia. During his training at the University of Berlin, he came under the influence of the new social ideas surfacing in Germany and subsequently led a group of émigrés to America in 1831. He established his reputation by applying a close knowledge of wire ropes, gained in the deep mines of central Europe, to make the first successful iron and then steel cable suspension bridges. Early projects included an aqueduct: the even load of water made an ideal application and allowed detail design issues, anchorages and hanger points to be worked up and resolved. By 1860 Roebling senior had graduated to the design of a railway bridge crossing below the Niagara falls. The problem of stiffening the structure, in this case a hybrid catenary and cable-stayed bridge, sufficient to distribute the concentrated load of the moving locomotive was tackled only as construction progressed.[4] By altering the stiffness of the deck girder relative to that of the cable superstructure and by adding tie-downs to impose stabilising loads, the structural behaviour of the bridge was made adequate. This manipulation gave invaluable insights into the behaviour of such systems. By the end of Roebling's career, his expertise and the scale of his schemes matched New York's ambition to bridge the East River. He did not live to see the Brooklyn bridge completed. Out surveying alignments for the project, his foot was crushed on disembarking. Forbidden the necessary amputation by his fierce beliefs, he succumbed to gangrene and it fell to his son, Washington (1837–1926), to carry on.

As Colonel Roebling of the Union army, Washington had learnt about temporary works on campaign, occasionally under fire. Early on in the conflict he set a suspension bridge across the Shenandoah river at Harper's Ferry, subsequently a major battlefield. He was in the observation balloon that first spotted Robert E. Lee manoeuvring towards the shoe factory at Gettysburg.

Completing his father's masterpiece involved mishap of his own. In order to make the mainspan practical, the pylons of the Brooklyn bridge had to be founded in the river bed on the shallow Manhattan bedrock. Foundation construction relied on a method perfected by the Frenchman Georges Triger whereby timber or metal cylinders, pressurised to keep the water out, would be sunk down to the riverbed then excavated onto a firm formation.[5] During a routine inspection Washington Roebling became trapped within the eastern pneumatic shell by a fire above. Caught for just half an hour longer than intended, he was crippled by the bends, in those days a misunderstood affliction known only from such work and termed 'caisson disease'.

## Emily Warren

Roebling's wife Emily, nee Warren (1843–1903), took over in her turn. Her husband assisted by observing operations from a wheelchair set inside a gazebo on nearby Telegraph Hill but it was Emily who fought the fierce day-to-day contract battles, spending evenings teaching herself the French structural theory upon which the design was based. Not until completion four years later and the hurly-burly done did Washington find himself able to walk again. In the meantime an unscrupulous contractor nearly ruined the project by supplying sub-standard wire. A complicated con using identical trucks circumvented Emily's testing regime. Rather than undo work completed, additional material was added to compensate. Today the bridge contains half as much wire again as was originally intended.

Emily Warren is the only woman engineer mentioned in this book; obviously the gender bias of engineering reflects historical and social conditions. Has this bias prevented a distinctively female contribution? In Emily's case her extraordinariness stemmed not from her assumption of a male imprinted role or from particularly feminine gifts of organisation and steadfastness. Instead it seems more to have been an event; gender proved irrelevant. In engineering, as increasingly it appears to be throughout design there may be no complementary masculine and feminine discourses. Engineering has no gender differentiation at all.

## The expansion of the railroads

Contemporary with the Roebling family's spectacular achievements, the expansion of the railroads southwards and westwards across the United States required many more prosaic medium-span crossings over the lazy, broad rivers of plain and delta. Combinations of factors determined the uptake of construction technology. Weak foundation conditions in the alluviums and loess of the Midwest precluded heavy abutments. There was a general under-capitalisation using untrained immigrant labour communicating in a second language, and the mills and foundries of the north were remote from the new sites. These factors led to the precipitate deployment of a particular bridge form. Metal trusses were standardised, comprising prefabricated components boxed and brought to the railhead for assembly. The development of the steel truss form and its transference between applications, countries and cultures shows many of the influences underlying technical change in those times.

## Trussed girder design: diversification

A secure patent system rewarded the North American technical entrepreneur and fuelled a well-documented transformation of truss design from timber patterns to iron and then steel.[6] Various compositions of strut and tie bear their inventors' names; several examples are pseudo-structures functioning despite structural misapprehensions and incoherent engineering thought. Patents by Wendel Bollmann (1814–1884),[7] Squire Whipple (1804–1888) and others embody conceptually simple hierarchical systems. Grounded in carpentry craft knowledge, these systems were enlarged through practical experiment. The variations on design themes were controlled by their amenability to graphical or elementary algebraic analysis of internal forces. Assemblages of stick-like parts readily idealise into understandable patterns of tension and compression travelling to the supports. Initially they allowed the size limitations of timber baulks to be overcome, then subsequently the use of cast and wrought metal components fabricated small enough to avoid serious flaws. A profusion of framework types and hybrids was applied indiscriminately to station canopies, train sheds and bridge girders.

There followed an extraordinary catalogue of disasters. Between 1880 and 1890 there were more than two hundred collapses;[8] design, manufacture and erection functions had drifted apart as processes depersonalised. Some bridges were simply too light, undersized in the competitive, unregulated market. Metalworking was an unpredictable business and unscrupulous contractors would press on with substandard products. Delivered in crates to the railhead, perhaps thousands of miles from their origin, these fully prefabricated systems were often put together by completely unskilled, illiterate labour. Any missing parts, perhaps highly stressed pins, that were mislaid might be substituted with the work of the local blacksmith.

Fail-safes and methods of achieving structural robustness became popular topics of research. As well as by the implementation of control procedures and testing regimes, modern quality assurance relies on a particular approach to design. The early problems with industrialised building in North America could be considered from a critical distance by émigré Europeans. The continuing influx of intellectual and practical rigour from across the Atlantic maintained a standard in the commercial pell-mell of North America.

Among many others, Albert Fink (1827–1897) brought his training in structural theory from the Berlin Technical Institute and applied it to the problems proliferating in truss bridge forms. Within a decade of coming to North America in 1870, he had risen to become the President of the Baltimore and Ohio Railroad.[9] His solution was foolproof, an utterly primitive hierarchy of simple truss elements: a large truss would be made up of two smaller ones further subdivided and so on (**figure 8.1**). The old additive sensibility of the mid-European carpenters was successfully

8.1
**Fink truss, Baltimore and Ohio Railway, 1860**
The different component sizes act as an instruction for the correct assembly of the whole. Each bay is a self-contained system, and load patterns in the structure remain balanced under the uneven loading of the rolling locomotive.

transferred to the metal fabricators of the New World. The assembly could only be made up in one way, with all the parts correctly positioned. Forces throughout the structure could be readily determined by graphics, vector diagrams of forces across each joint. For truss works the figures must close to be correct, and this property of 'self-checking' allowed safe analysis without calculation, all on a dusty drawing board at the construction site.

The 'Fink truss' was a successful stopgap, short-lived and rapidly improved upon. In all the variety, minimal weight and practicality of construction proved decisive criteria. So-called K-braced single cantilever span bridges came to predominate, and in the popular consciousness these large long-span bridge trusses have become evocative of the deep south of America – backgrounds in the stock footage of the freedom marches; the Tallahassee bridge of folk song. These structures belong to a specific time of expansion, a specific economic environment – one that was capital poor – and to a specific engineering consciousness in their design approach.

## James Eads

Mention has been made of Claude Navier's Seine bridge project destroyed by French backbiting (see page 107). Other malign political environments influenced engineering in North America. James Buchanan Eads (1820–1887) was an adventurer, beginning his career salvaging from the bottom of the Missouri river in St Louis. During the Civil War he turned to building ironclads for the Union navy, rapidly acquiring the craft knowledge essential to manipulate the early manifestations of iron and steel technology. With the peace and subsequent economic invasion, St Louis came to compete with Chicago as one of the two main jumping-off points for transportation westwards. Since its arrival there in 1840, the railway had been interrupted by the river with goods unloaded, transferred by ferry, and reloaded onwards. A bridge was essential to the city's prosperity. The river-boat lobby reacted predictably.

Eads combined his knowledge of the shifting sands of the river bed, which he had encountered first-hand from a diving bell, with his expertise in working iron to propose a beautiful, albeit expensive, arch bridge of unprecedented size and strength. Working outside the conceptual constraints of a contemporary engineering education, he schemed three long, low-trussed arches. His configuration used standardised steel components that could be replaced if necessary even after the structure was finished, and he drove construction through to completion against concerted and relentless opposition. Unnecessary clearances over the waterway were imposed by the ferry owners. An ingenious temporary works design overcame the restrictions (**figure 8.2**). He was vilified by the steel supplier Andrew Carnegie (1835–1919), who claimed that it would have been cheaper to build the structure out of silver. Eads tested every piece of metal Carnegie supplied in a special rig before installation, in the process establishing acceptance standards now universally adopted. Eventually the disruptions led him to enlist the help of his friend, General Ulysses Simpson Grant (1822–1885), who as President of the United States made a direct intervention to prevent the U S Army Corps of Engineers interfering in Eads's work.[10]

## Niche environments

North America's frontier mentality had generated a series of niche environments, places focusing extraordinary combinations of outlook, procurement possibility and functional requirement: design without precedent. In the Southern states a new form of river vessel appeared. Adapted to wide, sluggish and shoaling waterways and deltas, flat-bottomed punts were piled high with elaborate superstructures. Deckhouse walls were fretted for cross ventilation and shaded by deep verandas all around. Propulsion was by paddle wheels on each side or behind, robust and manoeuvrable in the shallows without any need for delicate rudders. Steamboat power plants had to be compact, and their design became a site for prolific experimentation and ferment of ideas. Hull strength came from elementary trussing, 'hog frames' perhaps alluding to ancient Egyptian precedent – cables strained up over the superstructure to support bow and stern.[11] Sometimes craneage would be incorporated into these girdles for cargo-handling or to lift paddle-wheels clear of obstructions.

## Innocent engineering

As well as being impressions of their environment, the forms of these vessels owed much to the background of their designers. Robert Fulton (1765–1815), an early pioneer of steamboats, made well-proportioned compositions whose elegance somehow managed to neutralise the ungainly agglomeration of disparate parts.[12] His former career as a miniaturist painter left his marine designs clear of received knowledge and hidebound technical expertise.

Wartime conditions force technical extremity, particularly onto the under-dog, and Fulton experimented with early submarines and torpedoes to attack the British in America's struggle for independence. By the outbreak of the Civil War, improvements in pressure vessel design made submersibles practicable. The Confederate submarine *Hunley* was a converted boiler, faired into an astonishingly prescient form.[13]

8.2
**Eads bridge,
St Louis, Missouri,
1867–1874**
Steel arches are fabricated incrementally with temporary support from an inverted fink truss. Air draft on the river is maintained and the simple suspension system could be adjusted with charcoal or ice to ensure that the two halves met perfectly.

As economic conditions worldwide diverged in the face of industrialisation, so local variations in engineering approach and attitude hardened into national characteristics. In North America the most obvious manifestation of a step change in outlook was a jarring increase in scale – a gigantism appearing not just in civil engineering structures, bridges and dams, but in all forms of building structure. A new kind of professional, the entrepreneurial architect, proposed tall buildings, sprawling factories, meat-packing plants and huge auditoria.

## Skyscrapers

Typifying the combination of businessman and fluent designer, Daniel Hudson Burnham (1846–1912) travelled to study at the French École des Beaux Arts. He returned to Chicago just as the indigenous form of the early settlers, timber shed construction, was being appropriated for the first skyscraper frames, tens of storeys high.[14]

A plentiful supply of timber, readily reduced and transported from the lumber mills back east, together with a climate limiting rot, made wooden buildings, trussed structures and light boarded balloon-frame shells and boxes the building staple of the Great Plains. The West was won using sheds thrown up very rapidly in the flimsiest possible framing of standard studs set at very close centres.

Together with other Chicago pioneers, Burnham took this typology and transposed it into frames of iron and then steel, using the rolled rails perfected for the permanent way.[15] Massive masonry buildings like the brick Monadnock Tower are proof that a propensity to build high preceded the technology, but the resolution of metal framing made it possible to build to heights previously inconceivable. In 1880 ten storeys were extreme; a decade later the Carson Pirie Scott store reached twenty storeys. Fireproof construction, telegraphic communication, controlled land speculation and the invention of the electric elevator were all attendant developments both resulting from and simultaneously contributing to increases in building size.

Over a long career, working with great flamboyance and imagination, Burnham established the principal characteristics of the new form and almost all its possible treatments. Concentrating structure into a load-bearing skeleton allowed the envelope of the building to be developed in many ways. The trajectory was not simply from massive enclosure to lightweight curtain wall. The façades of Burnham's Reliance Building, Chicago (1890), dissolve almost entirely into glass (**figure 8.3**). Ten years later his Flat Iron building in New York, the first structure to reach 100 feet high, has heavily rusticated quoins and cornices all suspended from a trabeated frame of riveted steel.

## Acquired engineering

Elsewhere processes of industrialisation and the appropriation of technology were appearing in even more compressed form. The islands of Japan, marginalised by a deliberate isolationism, completely reconfigured a feudal society around technological advance. The rush to modernise followed the shock of the gunboat diplomacy of American Matthew Perry (1794–1858) as a means to enforce trade. The changes were rather more than simple acquisition, a mimicking of western know-how. This palliative

8.3
**Reliance Building, Chicago, 1890–1895**
At 14 storeys, the modern office building seemed to spring into existence fully formed. The frame is riveted steel rails in a close-spaced 'balloon frame' configuration. The floors are fireproof tile and the cladding dissolves into a plate-glass screen with attenuated framing.

explanation invented by the West's self-image obscures the country's deep propensity for invention and improvement. Seemingly complete technologies were dismantled by subtle minds, reconfigured and extended far beyond their original intended range.

## Alexandre Eiffel

The European response to these world developments held on to a culture of economy and efficient material use, remaining conscious always of an aesthetic of engineering. The most famous name of the time, Frenchman Alexandre Gustave Eiffel (1832–1923), was a very practical engineer, fully exploiting new structural theory and applying rather than extending ideas. Hiring some of the best theoreticians of the day, his contracting firm bid for and won a series of design and build projects, steadily improving lightweight wrought iron structures for railway viaducts and termini. Large-scale bridge work gave him invaluable insights into the thermal movement of wrought iron. He concentrated on the development of specific structural forms, perfecting the lattice-braced strut, essential to increasing the scale of triangulated frameworks. Tenacious negotiating skills enabled him to deliver his eponymous tower for the Paris exhibition of 1889, commemorating the centenary of the Revolution (**figure 8.4**). In the face of entrenched reaction, he demonstrated a French and thence European commitment to technological ideology. The observation tower overtopped the previous tallest building, Chicago's Pirie Scott building, by fifty metres. The structure is massive but

8.4
**300-metre tower, Paris Exposition, 1887–1889**
Gustav Eiffel's tour de force: 7,300 tonnes structural
weight, 324 metres high.

acknowledges a structural ideal, tapering upwards into an 'eiffelised' profile. The form
was proposed and developed by a young assistant, Jacques Hubert (1830–1890).[16]

Eiffel's contribution to structural design did not, however, centre on ever-
increasing size. His prefabricated and beautifully rationalised train sheds, shopping
arcades and market halls, metal filigrees supporting glass and lightweight panels, were
readily accepted, even encouraged, by the sensibilities of French urbanity. These
buildings, dispatched to the outposts of French influence, defined new spaces, light-filled
volumes to which *fin-de-siècle* artists immediately responded (**figure 8.5**). Eventually

8.5

**Gare St Lazare. Painting by Claude-Oscar Monet
(1840–1926), 1877**
Monet's studies of light on the French coast around
Honfleur depict coruscation in the solid air. The
technique was transferred to this rendering of the
Normandy train returning to Paris in a smoke-filled
shed of metal bars and glass.

Eiffel was consumed by his milieu. Charged with corruption in the scandal surrounding Ferdinand de Lesseps's (1805–1894) failed Panama Canal project, he withdrew from commercial life, spending his last years studying aerodynamics, ever inquisitive.[17]

## Vladimir Suchov

In Russia protracted economic stagnation and social upheavals hampered building development. Coupled with an inferiority complex towards things western, this inactivity encouraged extremely high standards of theoretical thought. Engineers expected to apply 'first principles' to produce new form, and the work of Vladimir Suchov (1853–1939) embodies the idealism of the times. As a Ukrainian operating outside the mainstream of European understanding and with a long career spanning the revolutions, he remaining focused on structures for a new world. Surviving Stalin's earlier purges, he pursued pure form through the successive steel famines induced by centralised planning. Gaining experience alongside construction workers and fabricators, he applied a deeply theoretical grounding to produce metal structures of extreme lightness and originality.[18] New structural forms were required for the modern world and its new technologies; wireless transmission towers and shopping arcades like Moscow's famous Gum were unprecedented. Suchov experimented across the range: bridges, grid shell roofs and watertowers. His 'gittermasts', attenuated hyperbolic paraboloids, were true minimum weight forms (**figure 8.6**).

8.6
**Transmission towers, Oka river, Nizhny Novgorod, 1927–1929**
Vladimir Suchov built a number of industrial towers over 120 metres high using less than 2 tonnes of steel per metre of height. His 1919 proposal for a 350-metre tower required only 2,200 tonnes of steel.

The Russian urge to modernise (well recorded in Leo Tolstoy's writings),[19] proved important to the reform of older craft-based design traditions and to naval architecture particularly. Alexei Nickolaevich Krylov (1863–1945) addressed new scales of ship construction in an orderly way, combining theory, experiment and observation; hull shapes were to be determined hydro-dynamically and hull structures were to be efficient distributions of material.

## The aesthetics of technology in Germany

Despite such widespread acceptance, the belief in salvation through technical means seems to have peaked in the 1870s. Especially in Germany, the aesthetics of technology became a discussion topic in high schools and technical journals, and books and periodicals disseminated theory, codified practice and attempted to unify disparate outlooks. The way in which steel has been used almost to the present day was set at this time. The development of metal construction in building continued as a conversion of earlier timber and stone framing methods to the new material.

Meanwhile, the burgeoning industrial base was producing some very heavy engineering, marine engines, locomotives and gun barrels in quite unprecedented ways. Assemblies and components needed lifting around and out of factories. Rudolph Bredt (1842–1900) transformed crane design on behalf of the Mannesmann company, applying the latest theories of elasticity to the well-defined problems of crane construction, structurally scaled mechanisms built to very high specifications. The use of reliable high-strength steel produced slim components susceptible to buckling, and subject to secondary forces from imperfect joints; torsional stresses – twists – became important. Bredt and his colleagues found solutions working in an intellectual environment that close-coupled academy and industry. Results from these analyses of readily idealisable lifting equipment were first extended to the large sheds and enclosures needed to house this equipment, and thence into general building practice.[20]

## Teachers and researchers

Across Europe a new generation of engineering design teachers established itself. By completing systems, debating and publishing textbooks, this group fashioned the new theories to be amenable to practising engineers. At the ETH (Technical hochschule) in Zurich, Carl Culmann (1821–1881) and his follower Wilhelm Ritter (1847–1906) created a firm base from which to innovate. They maintained and improved an interest in traditional graphical analysis methods, at the same time professing an aesthetic of structural engineering.[21]

Other researchers were bringing on alternatives to these drawn and intuitive methods. A sequence of successes attracted talented mathematicians towards research in statics. Studies of differential equations seemed to show that certain of their characteristics, such as critical points, represented physical phenomena, providing otherwise unattainable insights into the nature of structural instability. Among a host of workers, D.J. Jourawski (1821–1891), Gustave Kirchoff (1824–1887), George Gabriel Stokes (1819–1903), Franz Neumann (1798–1895) and William John Macquorn Rankine

(1820–1872) collaborated and competed as they sought to systematise the theory of elasticity with ever more general results. They algebraicised earlier approaches based on geometry with the intention of discovering theoretical tools of universal applicability. Rankine was able to produce practical formulae for slender columns, soon tabulated for designers – the first steps to the automation of structural design.

## Alberto Castigliano

Parallel to this activity, the Italian mathematician Alberto Castigliano (1847–1884) was creating some of the most powerful tools deployed by today's engineers: the principle of virtual work, the associated energy methods of analysis and the notion of complementary energy. These are the most beautiful constructs of structural engineering. Applying an old principle, the generalisation of minimum energy, Castigliano concluded that a structure under load shakes itself down to a state embodying the least energy. How to determine this balanced condition? His brilliant nuance was to see that the response to any slight addition (a 'virtual' force) would reveal the structure's workings.

Developments of the idea, particularly applications to frameworks, were contested between the inventor and the Cambridge mathematician James Clerk Maxwell (1831–1879). Their competition was short lived and it was left to another Cambridge man, John William Strutt, Lord Rayleigh (1842–1919), to produce an amenable way of assessing the instability of lightweight frameworks. These theoretical advances were to be of immediate use in airframe design. The mechanical engineer Anthony Michell made another contribution, still coming to fruition, in a paper of 1904 by setting down elegant, simple but profound criteria for creating minimum weight frameworks – optimum structures.[22]

Emil Winkler (1835–1888) and Otto Mohr (1835–1918) developed elastic theory in three dimensions. A fourth axis, the development of time histories, encompassing problems of thermal movement, creep and shrinkage, was not far behind in their progression. Meanwhile, physicists and mathematicians like Heinrich Rudolph Hertz (1857–1894) followed up to resolve inconsistencies as they appeared, maintaining theoretical elasticity's wider research connections. Their generalisations were in turn re-appropriated and applied to specific engineering problems. The Pole Felix Jasinsky (1856–1899) transformed bridge truss design with his investigation of top boom stability. Understanding the tendency of these components to bow sideways made this common bridge form safe and efficient.

## Hardy Cross

The demand for textbooks made an industry out of academic engineering, standardising its products in the process. August Foppl (1854–1924) wrote several of the most widely read, influencing both designers and other authors. He ranged across a variety of structures, forcing a patchwork quilt of procedures into common format. Others explored specific problem areas in detail. Hardy Cross (1885–1959) is the most famous example. His method for frames is still taught in many schools; it is an odd thing, a numerical method of great beauty, startling power, and endearingly amenable to personalisation with its propensity for short cuts and deft resolutions.[23] Stephen Timoshenko (1878–1972), a pioneer of soil mechanics, the oddly underrated theory of manipulating the earth's surface, produced beautiful surveys of elasticity and structural stability, reconnecting these subjects with their historical origins.

## Science fiction and the use of speculation

Much was yet to come, and yet somehow a particular attitude towards engineering had been fixed. The translations by French poet Charles Baudelaire (1821–1867) of American poet Edgar Allan Poe's (1809–1849) speculations on science influenced Jules Verne (1825–1905), whose own explorations of technological initiatives became a literary genre, later termed science fiction.[24] Marvelling over the accuracy of Verne's predictions, the close approximation of his imaginary moon shot to the Apollo project one hundred years later, is to miss the point. His books were thought experiments, systematic and well-researched anticipations of the possible. The author consulted with contemporary engineers to calculate what a modern submarine would be like if an unlimited power source became available (**figure 8.7**). Ninety years after the publication of *Twenty Thousand Leagues Under the Sea* (its mysterious boat, the *Nautilus*, modelled on Robert Fulton's design),[25] the first atomic submarine, the American *Nautilus*, crossed the North Pole beneath the ice cap: literary invention and engineering innovation reconciled over one heroic period.

8.7
***H.L. Hunley*, 1863 (recovered 2004)**
The seven-tonne submersible propelled by hand crank killed thirty-two men during trials and one final successful attack on a Union sloop. Six years later Jules Verne imagined a 1,500-tonne submarine capable of uninterrupted circumnavigation.

Chapter 9

# Classical analysis and reinforced concrete (1890–1920)

Tracing the development of reinforced concrete leads one on a convoluted trail. The rapid rise to pre-eminence of the building material at the turn of last century is well documented,[1] yet repeated appropriations as a case study of technological evolution have left its history prone to simplification and misinterpretation.

## Pre-adaptations in the development of reinforced concrete

Artificial stone strengthened by metal bars arrived centre stage from a background of marginal applications and experiments. Proposals for improving the fire resistance of mass concrete by adding iron ties internally date back to 1830, with the underlying idea of preventing bursting under thermal stress. The notion of crack prevention then transferred itself to the more general case of controlling hydration shrinkage – the contraction that cement inevitably goes through when setting.

A cement boat reinforced with rods was exhibited at the Paris exposition of 1855; the material's water resistance was presented as its most useful characteristic. It seemed unlikely to replace more conventional hull build-ups, but in 1867, again in Paris, a gardener, Joseph Monier (1823–1906), matched the new material to a new use. Reacting to the Orientalism emanating from the colonies,[2] he successfully made and retailed a range of increasingly big pots for exotic flora in the new glass houses, verrières and conservatories of the bourgeoisie. These were cast from a concrete developed out of but not far removed from the Roman prototype, being bound with hydraulic cement rather than pozzolanic ashes. Wire reinforcement provided robustness. Having patented his 'system', Monier promptly began to speculate and experiment on the material's application to the making of floors, walls and arches in buildings.

Reinforced concrete has proved to be more than a man-made substitute for stone. Different approaches made early on have left a matrix of possibilities. Its usefulness is underpinned by several happy coincidences: the coefficients of thermal

9.0 *(opposite)*
**Notre Dame du Raincy, Paris, 1922**
The traditional typologies of church architecture are all retained. August Perret treats each component as an opportunity to demonstrate reinforced concrete's potential, even to the intricacies of tracery. Such prototypes most often contain more extreme applications than developed practice would allow.

expansion of concrete and iron are close together, so that the composite material does not pull itself apart when subject to temperature fluctuations; it is resilient to the harshest conditions. The fierce yet predictable shrinkage of the setting cement needs controlling by additions of aggregate, stone chips and sand, which also serve as cheap bulk fillers. The pulling together of the base material bonds reinforcement into acting in unison with its matrix: concrete literally squeezes itself onto the buried bars. The alkaline environment created by the chemistry of the cement protects the steel from rust. The thin bars are stabilised locally by the enveloping material and shielded from fire by the heat sink surround.

It was soon realised that just as shrinkage cracking could be distributed evenly through the material by wire reinforcement so also large-scale tension forces flowing from the configuration of the structure itself could be channelled into under-standable and manageable load paths passing through the elements. Compressive forces generate complementary tensions perpendicular to their line of action, and these bursting forces needed binding in. As bars could be laced up in arrays to closely match the patterns of forces, the use of strong but expensive steel could thus be minimised. And structural forms could be plastic, created in any shape that could be made of timber formwork by carpenters, wrights and joiners already expert in creating the flared hulls of ships and the furniture and cabinets of the Art Nouveau.

Despite concrete's hesitant inception, the clear economies available made it popular for commercial experimentation. It attracted a certain kind of engineer, one predisposed towards less confining conceptual environments, perhaps no longer offered by the wide open spaces abroad but found instead in Flaubert's or Zola's modern world.[3] Huge infrastructure projects, sewers, metros and roads demanded the new material. There emerged several alternative ways of treating concrete's structural action, discourses relating to reinforced concrete's mechanism and how it should be theorised and handled. One system eventually predominated. This process of selection and consolidation created its own framework of development. Some of the insights sidelined along the way are worth resuscitating. The work of four engineers, all French-speaking, illustrates the range of the enquiry.

## Engineering approach: presentation and accessibility

François Hennebique (1842–1921) won history's recognition as the principal pioneer of reinforced concrete construction. His commercial venture included the appropria-tion of a structural theory, a systematic development of detailing and a franchised method of construction that, marketed world-wide and exploiting French global influence, was responsible for the international uptake of monolithic concrete building. He attended to the meta-systems of reinforced concrete production, its standardisation, the communication of reinforcement systems through drawings and the conceptualisation of its structural action (**figure 9.1**).[4] Cheap semi-skilled labour maintains the predominance of the material in developing countries to this day.

Hennebique's pattern books offered a familiar kit of parts – the forms already received, divided and classified in the encyclopaedia of construction. His

development of reinforced concrete construction was inextricably linked to his marketing effort; the new material had to be presented in recognisable terms while at the same time differentiating itself from the conventional alternatives. The result was a quirky style of construction, and Hennebique's own house, a demonstration project, was a confection of cupolas, balconies and picture windows, conservative and outlandish at the same time. For a while he was one manufacturer among many, but thoroughness paid off and his method now dominates our thinking. Many of his contemporaries followed this path of re-casting conventional structural elements in the new material, but a few others set off on a wider-ranging search for new expression through its use.[5]

## Paul Cottancin

Paul Cottancin (1865–1928) patented his own system of reinforcing brickwork and concrete three years before Hennebique, one based on a novel conceptual approach which subsequently lapsed. He was born in Reims and the bureaucratic overtones of his French training were mitigated by his emigration to England, then Ireland. He combined contracting, engineering his own buildings, with consulting. As a researcher of reinforced concrete, he is a marginalised figure in the standard narratives of the material's development.[6] He arrived at a fully resolved alternative to the convention of seeing actions, tensions and compressions bunched up in bars and cement. His buildings comprised floors and walls that were conceived as completely isotropic

9.2
**Sainte Jean de Montmartre, Paris, 1904**
Paul Cottancin intuited a structural system of planes:
walls and floors acting compositely. The behaviour of
reinforced concrete was envisaged as a dissipation
of force, with fine mats of reinforcement rather than
heavy bars set within thin surfaces. His ideas suited
certain typologies: theatres and apartment blocks.

sheets of stress and strain,[7] his attitude to the new material being to regard it as 'simply' monolithic – not an unreasonable assumption – and then to work with the unity rather than the assemblage. Fine reinforcement meshes, nets and chain mail were distributed throughout the material, and he also tried out hollow masonry filled with concrete and laced with wires. Several of his surviving projects, churches and theatres, still display a unique and distinctive combination of arches, plate-like elements, struts and floors (**figure 9.2**).

## Pre-stressing concrete

Another radical thinker, one more established as a structural pioneer, was the precocious student and civil engineer Eugène Freyssinet (1879–1962), from the Limousin, again a graduate of the Écoles des Ponts et Chaussées. In his early twenties, building bridges on assignment far from home, he distracted himself with a close study of concrete's behaviour, particularly ways of addressing its inherent shrinkage problems.

Asked to implement three traditional masonry designs over the River Marne, he produced an alternative tender for low concrete arches, at a third of the competition's price, which was accepted by the government.[8] It was understood

9.3

**Airship hangar, Orly, Paris, 1917**
Eugene Freysinnet confronted a seemingly secondary issue in the use of reinforced concrete. In larger structures, bridges and sheds, shrinkage and long-term distortion under loading became significant. Pre-stressing, first with jacks, then with tendons, proved a solution of much more extensive application.

that the concrete would creep under stress and ongoing shrinkage would cause the arches to drop: in order to control this movement Freysinnet opened the arch rings at their crowns and then jacked them apart so that they became 'pre-stressed'. The introduction of forces to improve the distribution of stresses in a structure was not a new idea – we recall the bindings of primitive boats or the counterweighted pinnacles of Gothic buttressing – but the technique proved particularly useful in concrete structures, which not only shrink as they continue to harden over many years but also to creep very gradually away from applied forces.

Cold economics promoted the development of pre-stressing into a system for reducing section sizes by re-directing applied loads. As a side-effect the pre-compressions kept cracks closed, increasing longevity and improving waterproofing. The first experiments were with externally applied forces; then internal pre-stressing was tried by the introduction of tensioned bars within ducts (post-tensioning), or by the casting of concrete around stretched wires (pre-tensioning). The long-term shape changes in concrete tended to relax the tendons in ways difficult to predict, and the experience required restricted contractors entering the market. Freysinnet enjoyed a long and compelling lifetime's work developing practical applications of his insights. The pair of parabolic vaulted hangars he erected at Orly in 1923 were destroyed in the Second World War but the photographic record shows them to have been an astonishingly sophisticated alternative to the contemporary British and German projects **(figure 9.3)**.[9]

## Design vocabularies for new materials

The Belgian Auguste Perret (1874–1954) tapped into classical sensibilities in organising his own approach to reinforced concrete. By means of a contracting firm set up with his brother Gustav, he developed a particular form of framed construction, simple to construct but simultaneously seeking a real and appropriate expression of concrete's nature. Prototypes of shell roofs appeared from his office. This comprehensiveness, giving equal consideration to tectonics and architectural expression, led Perret to a 'vocabulary' of appropriate forms for the RC designer to compose with.[10] The block of flats at Rue Franklin, Paris (1903), displays panelled fields and skeleton construction while the church of Notre Dame du Raincy has a traditional layout transformed with attenuated members, pierced screens and cross vaults. Here is an interpretation of Viollet-le-Duc's structural rationalism (page 109).

## Initial over-refinement

As seems so often the case, the extreme attenuation and refinement of this early work is lost to much of modern practice. The exuberance generated by things intriguingly new, carefully studied and respected, retrenches into the over-familiar and received best practice. Over a single generation, reinforced concrete moved from new material to building industry staple, its use consolidated into a palette of constructability. But just at the moment that a single design approach established itself, so a reaction occurred as several important engineering figures began to revisit the boundaries of these methods and self-consciously sought other ways to build in concrete. These cycles of consolidation and innovation observable in engineering development are not always in phase between various places.

Such re-appraisals in approach mirrored the shifting and fragmentation of engineers' self-perceptions; different contexts and different problems faced broaden the variety of behavioural patterns at the same time that they condense specialities. The meteoric development of airframes is one locus to be discussed in due course; two others stem from a complex of attitudes arising in the German-speaking lands. A group of structural engineers defined themselves, sensitive to their cultural status and clear in their own minds that the ever increasing scale of their products, bridges, dams and building frames represented significant cultural artefacts. In the same milieu a special kind of 'architectural engineer' appeared – not just an engineer with design flair, but rather one closely aligned with the architect and seeking to coordinate an engineering input with an integrated whole.

## Swiss engineers

In Switzerland the analytical contribution of the academics Carl Culmann and Wilhelm Ritter have been mentioned (page 131). In their work and developing attitudes, they also created a unique conceptual space for engineers. By retrieving and revolution-ising the old sensibility of approaching analysis graphically, their diagrams achieved great power to conjure patterns of force. They worked in the Technical College at Zurich, a peculiar institution that then and now confronted many of the complexities and layerings of modern engineering. The school was set up in conscious reaction to the Napoleonic encoding of technology going on across the border, self-consciously positioning itself between older, almost medieval sensibilities of construction and the hopes of new structural discoveries promised by the mathematical theorists.

## Robert Maillart

One of the College's graduates was the celebrated engineer Robert Maillart (1872–1940), who made perfect expression of the engineer's art.[11] He relied much on full-scale prototype testing, and many of his insights and developments came through the use of model testing, using brittle plaster constructions that clearly showed crack patterns as they formed. The open spandrel characterising many of his later bridges follows straight from observation. Splits in models showed him where he could do away with useless infills. His bridges in the high valleys of the Swiss

**9.4**
**Salginatobel Bridge, Graubunden, 1929–1930**
Over a long and varied career, Robert Maillart identified and developed a particular bridge aesthetic based on the transposition of graphical analysis methods to built form. Diagrams of internal forces, envelopes, became the shapes of intrados and spandrels. Close observation of cracking in plaster test models confirmed his intuitions.

Cantons are effortless, direct reflections of their environment. Cheap structures linking villages and paid for by subscription, the bridges with their shallow arches fully exploit the rock abutments of the precipices to be crossed. The arch over the Salgina brook erected in 1929–1930 is a transitional form, fixing the moment when the traditional ordinance of arch with superimposed spandrels changed to one of contiguous stiffened arcs (**figure 9.4**).

## Marginal and hybrid forms

Maillart was self-conscious about his position as an artist, reflecting on it often. He would visit the Salignatobel bridge and muse over the visual appearance of the finished arch profile. His was the first exhibition of an engineer's work at the New York Museum of Modern Art in 1956. Other Zurich School members recorded their own positions on the relationship of engineering to art and on the respective roles of architect and engineer in the project of building.

New engineering sensibilities were not confined to any one material or form, and there are several structural types – marginal designs responding to peculiar problems or environments – that repay study. Bascule bridges and other very large moving structures were developed to meet the concentrated urban scales of the East Coast and Lakeside cities of the United States, in which mature dockland environments became intertwined with huge transport infrastructures. The different origins and expertises, so-called design cultures, of their inventors were reflected in their ordinance and detailing; slewing bearings on swing bridges came from the armament industries. Conventional bridge designers such as Joseph Baerman Strauss (1870–1938) opted for lifting bridges on simple trunnions. William Scherzer (1858–1893) came up with the ingenious rolling bascule, working like a rocking chair, and various railway engineers improved on its wheel/track combinations.

Transporter bridges were a tight commercial solution to a niche problem (**figure 9.5**). The idea was patented by the French pupil of Eiffel, Ferdinand Arnodin (1845–1924), following his work on the first example which seemingly sprang fully formed from the imagination of the Basque engineer Alberto Palacio (1856–1939).[12] A native of Bilbao, Palacio had specialised in iron and glass construction; he had studied in Paris under Eiffel and eventually completed the Atocha railway station in Madrid. Transporter bridges addressed the problem posed by crossings dominated by water traffic, harbour mouths, canals and navigable rivers. Limiting imposed loads to what could be carried on the transporter car meant that superstructures could be made as light as possible. The enthralling characteristic of these structures is the variety of details drawn from first principles then worked into the designs.[13] It is as if the outlandishness of the concept created a space for experimentation.

## Émigré engineers

In North America, European émigrés continued to leaven a stodgy engineering. Some of the larger railroad spans were of astonishing substance and weight. Gustav Lindenthal (1850–1935), born into the Austro-Hungarian empire at Brunn, now Brno in the Czech republic, had trained in Vienna, Munich and Switzerland before coming to America in 1874. The spare refinement of his first three years' work there in Pittsburgh – such as the Smithfield Street bridge – was abandoned upon his move to New York, where he was to produce the massive Queensboro bridge and Hell Gate Arch.[14]

9.5
**Marseille transporter bridge, 1904–1905**
The controlled imposed load permits a structure of extreme delicacy, rigged and adjustable to maintain running tolerances. Ferdinand Arnodin's previous bridge at Bizerta, Tunisia (1898), was dismantled and moved to Brest in 1909.

Many very heavy steel bridges were completed in this period, though quite why material should have been used in such a profligate way is unclear. Design-imposed loads were assessed at impossibly huge levels; inordinate allowance was made for construction defects instead of improving monitoring procedures. Nevertheless, one particular disaster cemented the prevalent conservatism. The pace of change had led to dislocations between construction capability, design technique and organisational power. The Quebec bridge disaster of 1907, the most lethal civil engineering disaster ever, was an avoidable tragedy (**figure 9.6**). In almost all engineering catastrophes there were several contributory factors.[15] This scheme was so ill-conceived that collapse was inevitable given the construction method used. The design team was hopelessly under-resourced and the design and construction of the bridge were let as one responsibility. Harbingers of the incipient failure, an increasing misalignment of members, halted work; the suggestion that prefabricated elements had been received on site already bent allowed a debate to be opened, and work was resumed before the design could be re-assessed. Seventy-four men were lost in the ensuing collapse. The design and the construction of the replacement were taken out of American hands and split between two independent Canadian organisations.

## Large suspension bridges

The very biggest crossings could not cope with excessive dead weight, and ingenuity and invention regained their imperatives. Joseph Strauss, the bascule bridge innovator mentioned above (see page 142), made a fortune building his patent designs. Born in Cincinnati, he had always hoped to be a poet, and a certain career restlessness culminated in the building of the Golden Gate suspension bridge (designed in the late 1920s and constructed between 1933 and 1937) linking San Francisco and Orange County. It remained the longest crossing for twenty-seven

**9.6**
**Quebec bridge collapse, 1907**
The collapse during construction of the first Quebec bridge was the inevitable result of a design mistake. Signs of incipient failure, local plate-buckling, were reported by workers, but without systematic interpretation the telegram ordering a stop was only on its way when the structure went down. Individuals were blamed, but the scale of undertaking had far outstripped the capabilities of personal professional responsibility.

9.7
**The Golden Gate bridge, San Francisco, as conceived, 1921**
Joseph Strauss made his fortune constructing patent-design bascule bridges for proliferating port areas. At the beginning of his long struggle to build San Francisco's Golden Gate bridge, the unprecedented scale of the task allowed his pre-conceptions to generate some outlandish proposals, the reassurance of the familiar completely without logic.

years after its completion. Strauss rejected a consensus that the project was impossible. He enabled and then took the credit for the design work of engineers Charles Ellis (1876–1949) and Leon Moisseiff (1872–1943) and architect Irvin Morrow (1884–1952). Finally ousted from construction control only a year before completion, he died broken-hearted just after the bridge was opened. The engineering critic Eugene S. Ferguson (1916–2004) provides a perceptive analysis of the slow, circuitous route by which the final design, seemingly inevitable, released itself from the preconceptions limiting Strauss's vision (**figures 9.7**, **9.8**).[16]

One reaction to the structural profligacy rife in North America came from an ex-employee of Lindenthal's, formerly a pupil of Wilhelm Ritter's back in Zurich. Othmar Herrmann Ammann (1879–1965) realised that the big combined road and rail bridges that he had worked on in Lilienthal's office would be superannuated. Henry Ford, whose Model T cars and lorries carried the mass migrations of the depression, revolutionised industrial production by the application of 'Springfield armoury practice' – interchangeable parts – to car manufacture and re-centred the entire nature of transport systems in the process. Ammann snatched the initiative off his old boss by proposing lightweight bridges restricted to road transport,[17] but Lilienthal overreacted by taking an untenable stance advocating only massive heavyweight behemoths carrying rail and road together. His obstructiveness probably sped his former protégé on his way.

9.8 (opposite)
**The Golden Gate bridge as constructed, 1933–1937**
Linking San Francisco with the underdeveloped peninsula of Monteray, the Golden Gate as constructed is a masterpiece of finely proportioned simplicity. The architect Irving Morrow established the practice of adjusting and emphasising dimensions to achieve visual repose, implying that structural and aesthetic ideals need not coincide. Large deformations throughout construction meant that every stage had to be carefully considered, an intrinsic safety feature of the suspension bridge form.

Extrapolating 'small deflection theory' to turn big scale to advantage, Ammann produced the George Washington suspension bridge (designed in 1923 and erected in 1927–1931), more than twice as long as preceding spans.[18] The design method took the established suspension cables/stiffening girder correlation and stood it on its head; the bigger the bridge became, the more its weight contributed to overall stiffness and the lighter the deck girder could be. Other engineers were slower to dispense with substance in their bigger designs. Trained in Paris, the Polish bridge designer Ralph Modjeski (1861–1940) achieved fame in the United States, reinstating simpler principles of brute strength and hefty deck stiffness in suspension bridge design.

## Very tall towers

The huge infrastructure projects of North America's 'New Deal' (1933–1938) and other political and economic responses to the global instability of the great depression promoted new organisational and procurement methods in civil engineering.[19] The economic impetus of very large projects had already been demonstrated in the construction of the Hoover Dam (1931–1935), a vast mass concrete plug between rock abutments on the Colorado river which had created an unprecedented logistical exercise in the Nevada wilderness.

At the same time that new bridges and roads appeared, urban skylines were being transformed by a succession of extreme skyscrapers such as the Woolworth Building and the Chrysler Building, culminating in the Empire State Building. The procurement base of this structure was so large as to generate its own momentum through the stagnating economy. Amid an abundance of resources, the design of multi-storey buildings became conservative within a very brief space of time. Frames were generally beams and columns, with rigid riveted joints and massive claddings precluding any dynamic response. They were expensive and ossified, their treatment over-determined for change.

## Structural indeterminacy

The analytical problem of structural indeterminacy, the assessment of interacting load paths, is particularly relevant to multi-storey frame design. A number of specific situations had been resolved over the years, but solution methods were now conclusively generalised. The academics' search for completeness in structural theory tended towards complex formulations; their drive was balanced by commercial pressure to find fast, fail-safe methods for repetitive conditions. Modern structural theory developed between these two poles.[20] All-encompassing abstractions paved the way for the automation of design at the same time that more and more pragmatic methods were being introduced to advance design in underdeveloped countries, spreading a western understanding of structural form.

## Classical statics

In England the notion of 'classical statics' maintained its hold. If structural problems were formulated in terms of the calculus, with general differential equations given specific boundary conditions relating to the real world, then, if solvable, the condition of all parts of a structure under all conditions could be exactly predicted. Professor Alfred John Sutton Pippard (1891–1969) at Imperial College, London, and Professor John Fleetwood Baker (1901–1985) of Cambridge University established a rigorous analytical technique which was adopted by English-speaking teaching institutions to take on worldwide influence. Their textbooks, dispersed across empire, persisted in print through many revisions.[21]

These two engineering mathematicians became involved in theorising the early aeronautical structures. The large-scale experiments and easily appreciable failure modes of such light frames influenced Pippard and Baker in different ways, and the dialectic between theory and experiment, described above between countries, departments and individuals, was to be played out within the career of John Baker. As Pippard went on to embellish classical theory, Baker made a new departure, one based on observation and which joined old notions of geometry with modern energy methods. 'Plastic theory' promised another revolution in structural design.

## Chapter 10

# Flight and the World Wars (1900–1950)

Structural design today bears the imprint of a periodic influence from the field of transportation engineering. Ship hulls, rigging, the development of rolled iron rails, rotating parts, chassis, road transportation all have structural requirements that have left their marks on the wider field, and the advent of aviation brought a completely new class of structure to the forefront of innovation.

10.0 *(opposite)*
**Vickers Wellington medium bomber airframes, 1936**
At the beginning of the Second World War, the almost overwhelming propaganda of German technical achievement was countered by a British emphasis on craft and design superiority. Barnes Wallis's design for a medium bomber dispersed structure evenly to avoid vulnerable concentrations of spars. The celebrated ability to absorb damage complicated construction; hand-built rather than production-line assembly.

## Flight achieved by compartmentalised solution

A host of pre-adaptations had to be drawn together to achieve powered flight. Sir George Cayley (1773–1857), a wealthy English landowner from Scarborough, had no technological base from which to explore his vision but, identifying the basic principles of flight, he produced a working glider and design for a powered machine, correctly predicting that power-plant weight would be the key to a practical machine. The Bavarian physicist Ludwig Prandtl (1875–1953) had been taught to closely observe nature by his father, a professor of engineering. He conceived of the boundary layer running across surfaces in the airstream and so put subsonic aerodynamics on a firm theoretical footing. Meanwhile, the Pomeranian Otto Lilienthal (1848–1896), a civil engineer and inventor, pioneered practical control techniques using manned gliders made with wings of very fine ribs shaping fabric surfaces,[1] eighteen experimental designs over five years resulting in a standard type. These gossamer structures relied on the interaction between the strength and stiffness of individual wing planes acting within a trussed biplane configuration. The delicacy of such assemblies required an almost impossibly fine line of design and Lilienthal perished as the result of a structural failure when a strong gust tipped his craft. The upper wing shed itself and the airframe lost all integrity.

## Process stability in design

The Wright brothers, Orville (1871–1948) and Wilbur (1867–1912), bicycle manufacturers from Dayton, Ohio, brought the diversity of research together in the design of their 'flyer', which first took to the air in December 1903. Both high-school dropouts, they taught themselves whatever skill was needed to fill the gaps in their project.[2] Their profession gave them the technical base and practical ability to minimise component weight as well as the resources to construct practical experimental wind tunnels and prototypes of workable power plant and control gear (**figure 10.1**). The sophistication of the Wrights' airplane lies not so much in its qualities as a good flying machine – its wing-warping control surface was quickly replaced by elevators and flaps – but in its practicality as a solution to a diversity of problems; not just light, but robust enough

to survive the inevitable setbacks, and readily demountable and transportable to the successive winter test campaigns at Kitty Hawk, North Carolina. The airframe amounted to a scaled-up version of a box kite, a 'canard' configuration, that had proved stable and strong in their preliminary tests. For their wing structure they chose a pratt truss configuration, timber strutting and wire bracing. The horizontal symmetry suited the strong reversals of loading at take-off and landing. Laminated ash ribs shaped muslin covers to form the aerofoils; the fabric weave was set at 45 degrees to the leading edge for stiffness. The wings were set off centre to compensate for the imbalance of pilot and engine side by side.

The Wrights made a relatively crude gasoline engine, the first of its kind with an aluminium crank case. It produced twelve horsepower when their plane needed eight. A pair of big, slow-moving propellers were fabricated of laminated spruce strengthened with varnished fabric, spinning in opposite directions to cancel gyroscopic forces and minimise loads on the airframe.

The concentration of handicraft evident in the first plane records the effort needed to hop over the threshold of powered flight for the first time. The 'design stability' of their approach allowed the Wrights to produce commercial aircraft within two years of that first flight. Other projects of the time were to prove more influential on the subsequent development of aviation, and patents and precedents were continually challenged. Their achievement was to recognise at a particular point in time that proven means would be just sufficient to achieve their objective. The time sapping and demoralising process of troubleshooting innovation, however important to the future vision of flight, was left to others.

## Beauty admitted to aeronautics

The focus on very lightweight components and systems in aeroplanes was to continue with important contributions coming from France. The painter Léon Levavasseur (1863–1922) made a fifty-horsepower V-engine, which became the mainstay of European aviation development up to the First World War. This reliable power plant provided a design centre around which structure and control could be developed. The flamboyant Brazilian Alberto Santos-Dumont (1873–1932) came to Paris as a student in 1891, and as the son of a coffee-plantation owner was able to finance a series of ballooning experiments. His response to the Wright brothers' achievement was to envision flight as a personal means of transport. He created the first 'ultra-light', the demoiselle monoplane, in 1907,[3] a structure of silk and bamboo weighing 235 pounds. The radiator was dispersed into tubes along the wing struts, the veins of the dragonfly, while a body harness warped the wings for control.

The war put a stop to this structural attenuation of aeroplane structures; the inertial forces generated by the high speeds and tight turns of aerial combat demanded different, stockier structures. Only in the design of lighter-than-air flying machines did the imperative of weight-saving persist, fostering an approach to lightweight structure made at the scale of building structures resisting relatively light and dispersed loads.

10.1 (*opposite*)
**Wright Flyer I,
1903**
The first generation of powered aircraft reflects its origins in bicycle technology. Wire-braced biplane wings are light and stiff enough to carry engine and pilot. The simple symmetrical pratt truss configuration of braces sustains load reversals on landings and steep turns. Components were beautifully crafted and fully demountable for transportation to the testing ground of Kitty Hawk. The sophisticated control system of wing-warping by adjusting braces was quickly replaced by simpler, more manageable hinged ailerons.

## Airship structures

So-called dirigible (controllable) airships had been imagined since the eighteenth century.[4] The French army general Jean Baptiste Marie Meusnier (1754–1793) produced a scheme as early as 1784 for an oval gas envelope supporting a powered gondola with propellers and rudder. A fish-shape design by the Swiss gunmaker Johannes Samuel Pauly (1766–1817), inventor of the firing pin, was tested in 1804 but not until 1852 did a manned flight succeed, an extremely dangerous voyage using a three-horsepower steam engine. The hydrogen-filled balloon managed sixteen miles, piloted by its designer Henri Giffard (1825–1882), a downward-facing funnel directing boiler sparks away from the envelope. Further development awaited the invention of the gasoline engine in 1896, and Santos-Dumont's work on control systems. Using a De Dion engine taken from his powered tricycle, Santos-Dumont followed a development programme through fourteen prototype airships and in 1901 he became world famous by winning the Deutsche de la Meurthe prize for circling the Eiffel Tower. This aviation challenge was also being pursued by Ferdinand Graf von Zeppelin (1838–1917), a German Junker and inventor with no doubts about the military applications of his chosen technology. As a military observer sent to the American Civil War, he had seen the Union army use manned balloons for artillery-spotting.

The first airships were structured as shaped membrane envelopes made rigid by their gas filling; a suspended fuselage and cable nets added stiffness. Zeppelin took the step of adding a rigid frame within the balloon and dividing it into a series of gas bags with cross frames in between. These compartments reduced the consequences of gas loss.

The apparent potential of the rigid dirigible – as intercontinental passenger liner and long-range bomber – won Zeppelin the backing of the German army and enabled him to persevere against many setbacks. Structural design issues centred on the overall stability of the skeletal frame assemblies, global buckling failure and the weight economy of individual elements. The braced strut was a key component appropriated from the iron and steel bridge designers and refined into tough minimum weight members. Compression elements had proved particularly amenable to mathematical analysis so that the 'ideal' strut was an identifiable objective that could be worked towards. Aluminium alloys were experimented with and their character-istics enhanced, and struts were pressed up out of plate with weight-reducing cut-outs. Riveted joints were progressively improved. Elements large and small were scrutinised for structural efficiency.

The development trajectory of the rigid airship spanned just forty years. Zeppelin's first aircraft, assembled in 1898, weighed twelve tonnes; the *Hindenburg*, the largest balloon ever built at 220 tonnes, closed the era when its hydrogen filling exploded at its New York landing point in 1937. In between there was a catalogue of failures, many of them structural collapses. Junctions between fuselage and fin, in particular, caused insurmountable stress concentrations, disturbing otherwise dispersed and flexible systems. Even as airships grew to enormous sizes their use remained questionable.

Despite early success as a terror weapon in the First World War, Zeppelins proved very vulnerable and by 1917 were withdrawn to transport duties only. The British initiative in airship construction received a major setback with the R101 disaster in 1930, when their flagship hit a hillside near Beauvais in France on her maiden flight to India.[5] Against strong senate opposition the United States navy persevered with an airship programme, trying to meet the strategic problem of patrolling the country's western approaches, and with a virtual monopoly in helium production, a gas giving less lift than hydrogen but which is far less combustible, the North Americans created aerial aircraft carriers scouting across the empty spaces of the Pacific. Biplanes would drop from internal hangars to search beyond the horizon and return along a pre-arranged pattern to hook onto the mother ship.[6] If the Navy's two principal craft, the USS *Akron* and USS *Macon* had not been lost – the former during a storm off Big Sur in 1933, the latter to upper fin collapse in 1935 – then their ability to scan vast spaces might have helped prevent surprise attack.

That a development momentum was maintained in the face of these disasters is intriguing. Propaganda was important; symptoms of nationalism at the turn of the century included not only arms build-ups but other technological races as well. The transatlantic flight of the *Hindenburg* was promoted by the Nazi regime, inveterate risk takers, specifically as a demonstration after the American setbacks. Airships symbolised a certain dream of modernism coupled to transport technology and attenuated structure, and regularly appeared in the renderings of Russian 'Constructivist' architects. Their threat prompted the bomb-hardened roofs of Le Corbusier's projects for a new urbanism in the 1930s.[7]

The undoubted idealism of the early airship designers and their 'accelerated environment' led to over-confidence and hubris. Step changes in scale made almost overnight simply magnified otherwise minor structural effects. There was a propensity to take on major design changes at the drop of a hat. When it was decided to make the R101 the mail boat to India, it was realised that gasoline fuel might prove volatile in the heat. A switch to diesel power solved that issue but increased weight dramatically; the solution was to saw the ship in half and add another segment. The airframe was checked for adequacy and strengthened. It was concluded at the crash enquiry that a ludicrously naïve detail – padding to prevent chafing between gas bag and rib – had not prevented leakage. Calculations and details in support of a design variation, made under time pressure, seldom have the same assiduity as the original scheming.

## Military development: acceleration and atrophy

The single extended conflict that became the two world wars and cold war controlled the entire first half of the twentieth century. With very little diversion the development of flight focused on creating war machines, and air frames were shaped by their importance to the arms race. The largest passenger airliner in regular use today is a modified bomber design.

The First World War fixed a particular configuration of aeroplane and aircraft structure. Better controls and robustness were essential. Planes were being used in the worst of conditions, operated from makeshift airfields, and artillery-spotting, reliability and stability in the air quickly ceded prime function to combat for control of airspace. Dogfighting became the characteristic aerial confrontation, so performance and substance to act as a weapons platform were the critical concerns. At the same time that there was a diversification of types – experiments to gain the upper hand – there was a convergence on small, simply configured scouts (**figure 10.2**).

Such planes could make the most of rapidly increasing engine power to dive and climb. Shorter wings turned away better and were less prone to warp or shed under the inertial forces of extreme manoeuvres.[8] Timber and fabric construction was taken to the limit of its practicality. While fighter plane designers looked elsewhere for structural forms to gain competitive advantage, canvas-covered

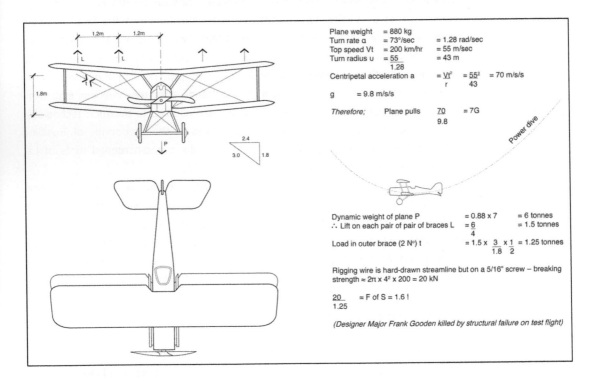

Plane weight $= 880$ kg
Turn rate $\alpha$ $= 73°/sec$ $= 1.28$ rad/sec
Top speed Vt $= 200$ km/hr $= 55$ m/sec
Turn radius $\upsilon$ $= \dfrac{55}{1.28}$ $= 43$ m
Centripetal acceleration $a$ $= \dfrac{Vt^2}{r} = \dfrac{55^2}{43} = 70$ m/s/s

$g$ $= 9.8$ m/s/s

*Therefore;* Plane pulls $\dfrac{70}{9.8}$ $= 7G$

Power dive

Dynamic weight of plane P $= 0.88 \times 7$ $= 6$ tonnes
$\therefore$ Lift on each pair of pair of braces L $= \dfrac{6}{4}$ $= 1.5$ tonnes
Load in outer brace (2 N°) t $= 1.5 \times \dfrac{3}{1.8} \times \dfrac{1}{2} = 1.25$ tonnes

Rigging wire is hard-drawn streamline but on a 5/16" screw – breaking strength $\approx 2\pi \times 4^2 \times 200 = 20$ kN

$\dfrac{20}{1.25}$ $= F$ of $S = 1.6$ !

*(Designer Major Frank Gooden killed by structural failure on test flight)*

**10.2**

**Royal Aircraft Factory SE5a, 1917**

The air battle over the Western Front of the First World War escalated quickly into a scramble for speed and robustness, fast platforms, tight turning and resistance to the vibration and recoil of heavy machine guns. Stocky airframes carried compact power plant, and heavy wings were increasingly frame-braced rather than wire-rigged.

The completion of a particularly powerful engine, the inline Hispano Suiza, brought 20 per cent more energy than rivals in a rather unwieldy and extended block. A strong simple fighter plane was designed around it to become one of the most successful weapons of the later conflict.

**10.3**
**Vickers Vimy**
**heavy bomber,**
**1917**
The design of
the first plane
across the Atlantic
concluded
early lines of
development.
Effectively
proven systems
of construction,
wing bracing,
undercarriage, etc.,
were multiplied
and extrapolated,
making the plane's
use as bomber then
passenger carrier
very short-lived.
Alcock and Brown's
exploit crossing the
ocean was typical
of contemporary
adventures
over-extending
equipment use.

frames persisted in the heavier conservative designs of bombers. A smattering of
bombs had been dropped by the early Zeppelins, but it was not until the closing
phase of the First War that production was allocated to bomber planes as their potential was recognised. The relationship between aerial bombardment, transcontinental
flight and global communication was quickly established. Alcock and Brown made it
– just – across the Atlantic in a converted Vimy bomber, as the Versailles treaty was
about to be signed. Their plane was constructed out of a hierarchy of simpler systems
collaged together in the way that primitive structures are always first made large. The
design was modified to become a rather romantic passenger airliner, with a wicker-
seated cabin and the pilot in an open cockpit above, plying the British overland routes
across Africa and eastwards as far as Australia. The ungainly framework, adaptable
and patchable, represented the end of a line of development (**figure 10.3**).

## Over-extension and pre-determination

A technology not yet fifteen years old had been stretched to its limit crossing oceans
and continents.[9] But already the war had brought on other alternatives, a period of
rapid, almost exuberant mutation from which a development path appears, then the
grim consolidation of that path. The German Junkers J1 plane was something quite
different from other lightweight braced frames (**figure 10.4**), the original intention
being to make a stout aircraft immune to ground fire. A covering of sheet steel to
wings and fuselage was stiffened with steel ribs. Flight surfaces and body were still
the independent components recognisable to the riggers. These men were setting
up and thereby continually assessing the early designs. In riveting the plates into

continuous tubes braced by a subframe the metal surfaces became stressed, and the whole structure reached a new level of strength and stiffness. Though slow to climb and turn, the ugly plane was among the fastest of its time and improvement was rapid; ribbing was reduced by corrugating or curving the plates, gaining strength and lightness simultaneously. Tactics changed: instead of wheeling and sparring, the first high-speed pass became decisive. Power and speed became the principal objectives of designers.

Came the peace, and competition continued through a series of prize challenges. The interwar Schneider trophy races, floatplanes over the sea where take-off and landing distances could be almost infinite, focused performance into pure speed on the turn.[10] These stubby over-wrought racers converted almost directly into the petrol-driven monoplane scouts of the Second World War. The Supermarine Spitfire was derived from a prize winner. It was well equipped, with elliptical wings to turn away faster than its assailant. A kind of technological see-saw set in; designers seeking lightness and manoeuvrability, then speed and strength. The advent of the jet engine at the war's close seemed to settle matters in favour of pure power. Not until the Korean War and the success of Russia's MIG 15 jet did manoeuvrability and rate of climb gain unexpected advantage again.

## Idiosyncratic engineering

The technological scramble of the war left room for the quirkiness of designers to flourish. In Britain the measured contributions of Pippard and Baker have been mentioned (page 147). In counterpoint to their work came the career of the eminently

10.4
**Junkers J1, Blechesel (sheet metal donkey), 1915**
Small-arms ground fire brought down many observation planes in the First World War. In the search for robustness, stressed metal skins, fuselage and wings proved the crucial pre-adaptation for plane development. The first monocoque planes were slow, vulnerable to fleeter scouts, but their strength duly transformed engine powers and aerobatic potentials.

British Sir Barnes Neville Wallis (1887–1979), who fitted himself into a tradition of idiosyncratic engineering excellence.[11] He came to aviation from a background in shipbuilding, joining the airship design team at the manufacturer Vickers Aircraft. The fate of the R101, one of the two major airships commissioned by the British government between the wars to its own design, has been described, but the R100, with a geodesic (triangulated) braced frame engineered by Wallis, successfully made the round trip from England to Canada. The structure proved adequate, but protecting the gas bag impossible with the technology available, and the ship was broken up.

Wallis and his ideas went forward. The geodesic frame principle was applied to the design of a medium-range bomber that became the famous Wellington. The aircraft's performance was uninspiring, but the multiple load paths through the diamond-pattern ribbing made it uniquely resistant to flak damage. However, the number of joints made it expensive to build – a quixotic strand of profligacy that can be detected elsewhere in the British wartime design effort, like a subconscious drive to win only with style. The simple brake-pressed lines of the Messerschmitt fighter had the martial practicality of that first Junker monocoque. The Spitfire was compounded of elegant double curved surfaces. Did the British aircraft really take twice as long to make as its German counterpart and, if so, was that extra effort reassurance of fine margins of superiority? The American Packard company quickly simplified the Rolls-Royce Merlin engine for mass production, gaining weight and losing tone but not real performance. German designers always seemed to have taken more account of production problems, finding a different balance of best equipment against best use of resources.

Barnes Wallis maintained his pursuit of the extreme in ordnance design, creating inordinately big conventional weapons. The bouncing bomb is legendary. The destruction of three Ruhr dams in 1944 was achieved at great cost but tapped into a promise of technology at a time when morale was low. The British nation accepted that gifted and inspired individualism could surmount seemingly extraordinary obstacles: the maverick engineer was a national hero.

## Weight-saving in very large structures

So the war years had produced an industrial base capable of building the very lightest structures. Correspondingly, competition for transatlantic trade and then the arms build-up towards the First World War generated ships of unprecedented scale. Vessel dimensions tripled during the dreadnought race,[12] and despite displacements upwards of 50,000 tonnes, the structural theory applied to hull design remained relatively simple. This conservatism is surprising. Improvement in guns made armour-carrying capacity crucial, and weight-saving was imperative. The very size of the vessels made the unpredictability of wave loading relatively insignificant so that designers could concentrate on achieving the maximum stiffness and strength from conventional configurations of keel and ribs. Hulls were effectively long beams, loaded by engines and superstructures amidships and having to span and cantilever across the waves.

The Japanese navy, seemingly less constrained by long tradition, was quicker than its potential opponents at embracing new forms, and a generation of talented designers were given their head. Cruisers, the long-range scouts of the fleet, proved particularly good sites for structural innovation. They had to be fast and light yet carry enough armour to survive first contact. Substantial weight savings were available by dispensing with ribs and stringers. Hull and deck were reconceived as one large box beam. The Japanese 'A' class was an extreme design with one continuous sweep of deck from stem to stern, expensive to fabricate but directly reflecting the distribution of stresses along the ship's length.[13] Secondary components were also pressed into service as primary structure in this design; the swept funnel structures created a multi-cellular structure within the hull. The naval architect was Fujimoto Kikuo (1887–1935), a visionary and original individual who died prematurely following a court martial when one exploratory design too many failed. His prototype torpedo boat capsized.

## Artificial constraints forcing developments

Man-made constraints were to produce many of the more ingenious detail innovations in ship design. The treaty of Versailles, which ended the First World War, capped the size of the German navy's battleships, and there followed a sustained development programme to circumvent the rules. Results included significant advances in propulsion units, large diesel engines and steel welding. Germany's so-called 'pocket battleships' were intended as surface raiders, prowling the shipping lanes.[14] Oil-fired, they did not give off telltale puffs of black smoke on the horizon from stoking up to catch their prey. Their major structural advance, one made much of in propaganda to establish German technical superiority, was in the joining of hull plates. Ships were generally constructed of steel sheets, staggered and riveted; butt joints would save the lap weight and so electric arc welding, a sensitive process, was refined to make the deep long seams required. The Allies in turn exploited this technology for their Liberty ships, standardised supply vessels making the Atlantic run. The large-scale use of welding speeded their construction, and structure was adjusted so that sectionalised segments could be prefabricated then brought together and run off the slipways in mass-produced sequence (**figure 10.5**).[15] This experience of making and shunting was applied in the post-war reconstruction. The designs of the box girder bridges of Germany and central Europe, paid for by the Marshall plan to replace the crossings destroyed in the final invasion of the war, are a legacy of this expertise.

The austerity of the war years promoted the use of reinforced concrete. Steel was precious, and the simple technology of mixing and casting earth materials allowed unskilled, replacement and slave labour to produce fortifications and production facilities. And the material was almost ideally suited to resisting

10.5
**S.S. Schenectady , Swan Island, Oregon. 1943.**
The mass production of shipping required to transport materials to Europe in the Second World War promoted the handling and joining of very large-scale components. Long continuous welds proved brittle as heat distortions failed to dissipate in thick plates flexed by the waves. Ships occasionally unzipped themselves in calm harbour conditions.

bombardment. Almost completely fireproof, reinforced concrete walls and floors yield away from blast; the matrix cracks and the bars deform but retain their integrity. Enormous energy-absorbing capacity was available to military engineers for bunkers and submarine pens.

## Austerity and production in-balances; the efficient and imaginative use of materials

John Baker was cited earlier (page 147) as an exemplary engineer of Britain's wartime establishment. His early years as a structural designer of airships came to an end when he was diagnosed tubercular at the age of twenty-eight, and he turned to research work for Britain's Structural Steel Research Committee, a government group formed to produce a practical design code for building frames. Full-size tests revealed that real stresses in steel structures bore little relation to those derived from elastic theory. Efforts to address this revelation coincided with directives from the Minister for Home Security calling for research to reduce the vulnerability of buildings to bomb attacks, which could be most practically achieved by making structures more ductile, so that they would bend rather than shatter under impacts.

## The plastic theory of structural design

Up to the time of this research, structures had been assessed in terms of the loads at which they would collapse rather than for the stresses to which they were subject in normal conditions. This 'limit state' approach pointed a way to designing safe, economic structures immune to oversights of real conditions. The war had shown that severely abused buildings could be readily replaced. If failure mechanisms were understood and adequately protected against – the so-called 'ultimate limit states' – then economy could be safely pursued at working conditions, designated 'serviceability limit states'.

By 1948 Baker had a practical method of designing multi-storey steel frames, sketching in plastic hinges that formed at collapse then calculating the work done to strain the structure to that point. The fundamental theorems of plasticity followed on from this work, extending its use into other forms of structures, particularly plates and slabs. Masonry structures, including existing arch bridges needing reassessment for military loadings, and also much of the country's medieval building heritage, proved particularly amenable to this method of analysis.

The introduction of plastic theory was momentous; no new theory had appeared in the field of engineering for a century.[16] The process involved recasting the behaviour of structures into real time instead of imagining them static and immutable. Acknowledgment is made of the way a structure undergoes a number of transitions on its way to failure, and it is the study of these transitions using energy principles that constitutes plastic theory. Geometry persists in the notions of plastic hinges, instantaneous centres of rotation, and the constraints that are imposed to maintain compatibility.

Across his career Baker had moved from the practical application of classical 'elastic' structural theory, through the invention of a practical design, too, concluding with its theorising; his last years were spent as an academic, propagating the new ideas. Intellectually attractive and simple in principle, the approach to analysis through an understanding of real behaviour, plastic theory dominated British undergraduate teaching through the 1960s. Elastic design has proved easier to computerise and the further development of plasticity soon stalled.

## Structures supporting new technologies

Occasionally unique forms of structure appeared associated with a new technology. As an outcome of experiments by Guglielmo Marconi (1874–1937), radio communication had become practical in the 1890s while the acronym radar (radio detection and ranging) was coined in 1941. In between, large-scale transmission masts were perfected. They differ from earlier rigged structures in their high level of flexibility and interaction between strut and stays, and as a set piece of structural design they attracted much research. Relatively low capital cost made prototype testing practical; the masts could be readily instrumented to measure applied loads, particularly wind, and to record structural response. New models of gust loadings were created from statistical data applicable to all other, more inert buildings. Last century, the refinement of loading assessments contributed far more to structural efficiency than the improvements of materials science.

## Over-sophistication in engineering

It would be a grave oversimplification to suggest that the wars only acted to accelerate technical advances in an atmosphere of desperate experimentation and aggressive innovation. A kind of design conservatism also perpetuated pre-existing traits in the various technologies and this effect was reinforced by pressure for fast implementation. The maintenance of indigenous characteristics sometimes proved important. The Germans invaded Russia with the most advanced armoured vehicles in the world at a time when the country's engineering base had been decimated by its leader, Joseph Stalin (1878–1953), so when the fight-back came tractor designers, already set up far to the east for internal colonisation and the industrialisation of farming, had to be pressed into service and it was they who were able to deliver a design of tank truly suited to their own ground and the people called upon to use it.[17]

Propaganda is by definition deceptive, and German technical advances were by no means smooth or continuous. In the battle of the Atlantic the choice was made to direct resources to the production of a First World war design of U-boat rather than research, building numbers instead of maintaining the technology gap against defence systems, but the gamble was lost, and the happy times turned into a glut of sitting ducks. It was not until the last desperate days that a lethal design of submarine burning hydrogen peroxide fuel was converted into a diesel electric hybrid, a serendipitous change that immediately created a completely new generation of underwater vessel.[18] Too late to affect the outcome, the new boat became the progenitor of all future development, the influence of its profile appearing in the silhouettes of the first nuclear vessels.

## Fracture mechanics

Pressure on resources had materials science improving its products in leaps and bounds – not just metals but also the earlier polymers and natural fibres. The concept of fracture mechanics had been introduced during the First World War by the aeronautical engineer Alan Arnold Griffith (1893–1963) to explain brittle failures.[19] The microscopic behaviour of materials, the way cracks propagated from surface flaws, led to the development of a particular palette of structural materials suited to severe conditions – elastic and ductile, with reserves of energy absorption when approaching failure point.

## Old materials, new ways

Plywood, timber veneers hot-pressed together with a Bakelite bonding, proved sufficiently durable to exploit the excellent strength-to-weight ratio of cellulose fibre. The celebrated Mosquito light bomber outpaced all pursuit supported by a timber airframe, and when political opposition denied the Hughes Corporation aluminium supplies to build its design for a huge troop-carrying sea plane, Howard Hughes (1905–1976) switched to plywood stressed skin construction throughout fuselage and wings.[20] The so-called Spruce Goose was completely compromised by control problems improperly addressed at the new scale, but the plane proved itself strong enough to lift off. Composites were on their way to being practical materials for the very largest and most highly stressed structures.

## Transmuting manufacturing capacity and its products

Came the peace, and manufacturers of long-range bombers switched their output towards transcontinental travel, and North America's interest in Europe and expanding internal trade relied on regular long-distance flying. As the cold war set in, the shortest route to confrontation with thermonuclear weapons lay along great circles across the North Pole. The unrelenting distances involved required airframes of yet bigger scale, delivering heavy weapons at great range. The American B52, originally required to carry a 10,000 pound payload a distance of 5,000 miles, was effectively designed, in response to a government competition, in a suite of hotel rooms over a single weekend.[21] The form of the plane was almost completely conventional and went on to form the basis of Boeing's most successful civilian design, the 'Jumbo jet'.[22] Lightweight metal and composite technologies developed for the aircraft industry provided a ready-made platform from which to develop first ballistic missiles and then space vehicles. Even these extreme structures took their form more on aviation precedent than on first principle design and only now, half a century later, are alternatives to this manufacturing base gaining real credibility.

## Flaws in the conceptual basis of structural engineering

By the middle of the twentieth century, the high technologies seemed to carry all of the future's promise. It was accepted that reconstruction required new means, and all development was welcomed as never-ending. The philosophical foundations propping up this belief were diverse, but that the products of modernism were flawed was obvious. The war effort had culminated in the possibility of humanity's complete self-destruction, and the social changes concurrent with the reconstruction failed to bring universal happiness.

In construction, output rose but better environments remained elusive. There was a renewed interest in form, the underlying structure of things that gave the world coherence, and various thinkers were influenced by concepts taken from engineering and construction.[23] The celebrated Austrian philosopher Ludwig Wittgenstein (1889–1951) trained as a mechanical engineer at the Technische Hochschule in Charlottenburg, Berlin. His interest in aeronautics – powered flight was then just three years old – was marred by dissatisfaction with the theoretic under-pinnings of the many concepts involved and he began to note down philosophical questions, reflecting on the concept of force and whether it had any meaning beyond its function as a mental construct.[24] In 1908 he went to Manchester to study aeroplane design before his misgivings had him directed to the Department of Philosophy at Cambridge University. As soldier, then prisoner of war, he delved into the logical basis of mathematics, working towards a system that made out scientific and technical concepts to be purely synthetic. Our ideas on engineering are as much artefacts as the structures we make.

Another German-speaking philosopher, Martin Heidegger (1889–1976), developed his theory of being – what it means to exist – by theorising a 'philosophy of building'. Amid the destruction of the Second World War, his thinking became less systematic. He found nihilism at the base of modern technological society yet at the same time situated the process of construction, of making things appropriately, within the essence of being.[25]

Chapter 11

# Early contemporaries (1945–1960)

## Post-war reappraisal of conventions

The upheavals of the two world wars shattered intellectual barriers. There had been continuous and precipitate changes, and globalisation had been allowed to take hold. High cultures had proved base, with others mixed beyond recognition. German Jewish thought had been scattered westwards while the Americans overran Europe.[1] The British Empire collapsed and the superpowers ascended and opposed one another. The chaos represented opportunity, and all forms of rebuilding were sites for new thought. Technology was initially held to be the route to salvation.

Despite this assumption now coming under some pressure, we nevertheless remain within the event that occurred and the aftershocks will persist. The directions taken and ideas adopted in the four decades following the Second World War, when pragmatic engineering became veneered with a new sensibility, are the ones we are currently working through.

A new self-consciousness among engineers revealed itself as a belief that reliance on technology amounted to a culture in itself. Engineers were not trying to be designers per se but began to pick and choose between the emphases they would apply to their work. They were not seeking to mimic architects but joined in the re-emphasis on underlying form and the response to machine production and prefabrication that was taking place.

## Individuality: Felix Candela

Within all the turmoil, the responses and experiences of individuals were to prove more important than more general currents of consensus. There were some extraordinary personalities, among whom Felix Candela (1910–1997) is exemplary. A Spaniard of republican sensibility, he narrowly survived the civil war years (1936–1939), imprisoned in Perpignan. He spent the time studying mathematics, particularly calculus, and upon release sailed for Mexico. In Veracruz he set up a contracting firm with his brother, pioneering in the design and construction of concrete shell structures, and the combination of benign climate, cheap skilled labour and Latin-American Baroque flowered in pure architectural forms.[2] As a building

11.0 (opposite)
**La Manantiales restaurant, Xochimilco, Mexico,1958**
The almost hallucinogenic forms of Iberian baroque coincide with Felix Candela's investigation of double-curved surfaces. He made shapes definable by differential equations. Solved by consideration of the boundary conditions, these formulations produced elegant forms with light, feathered edges.

11.1
**Wyss Garden Centre, Solothurn, Switzerland, 1961**
The form-finding of Swiss engineer Heinz Isler involved the use of tension analogies: inverted models and soap bubbles. Strains under self-weight are minimised, ensuring efficient material use. The profiles have a natural repose. Edges tended towards substantial framing.

contractor, he coupled a craft understanding for his chosen material, reinforced concrete, with a deep regard for the elegance and simplicity of the differential equations and their solutions which govern surface structures.

Candela's facility for classical analysis attained to a guiding principle. At school he had made pocket money doing maths assignments for other students, and elegant solutions took on special significance for him. He held that appropriate forms were those fully describable by soluble differential equations; there was an underlying correlation between the physical grace and balance of a structure and the intellectual beauty of the suite of equations needed to describe and analyse it. Using minimal amounts of material, his roofs were composed of simple groupings and distortions of basic forms. From a limited vocabulary he made all kinds of buildings: a factory for Bacardi Rum in Mexico City; an open-air restaurant in Xochimico, its vaults reflecting the exotic foliage of surrounding gardens; the Church of the Miraculous Virgin in Mexico City, top-lit and sublime.

There is an instructive comparison to be made between Candela's work and that of Heinz Isler (1926–). The latter was born and stayed in Zollikon, now a suburb of Zurich, and his experiments with concrete vaults covered the same ground as Candela's but in a very different way and with different results (**figure 11.1**). Beautifully executed and attenuated to the extreme, Isler's shells are very much static exercises progressing towards increasingly pure and neutral forms. His designs were generated by working up models and not from well-turned solutions to differential equations.[3] One winter he sprayed draped netting with water to build up shells of rime. More usually his studies centred around tension analogues; chain nets hung in space took into three dimensions the old idea of inverting a hanging string to find an ideal arch profile.

## Frei Otto and the German reconstruction

For the Saxon Frei Otto (1925–), apprentice stonemason turned Luftwaffe pilot, the war ended in a Russian prison camp, and he relates how, together with other inmates, he got by on dreams of the reconstruction. His hope turned into a lifelong exploration of tented forms.[4] Lightweight fabric structures could cover large areas economically; their contingent appearance was acceptable to the times, and the early prototypes were pavilions in beer gardens and parks for a nation become used to living outside. The central European circus tradition provided tentmakers with the expertise to pattern and fabricate very large membrane structures, and the chemical-industrial base could produce the kind of veneered weaves from which permanent structures could be made. Working together with the German architect Günter Benisch (1922–), a survivor from the U-boats, Otto produced a tour de force demonstration of tensile roofing for the Munich Olympic Stadium of 1960. A vast cable net, gilded with scales of plexiglas, was stretched over seating banks heaped up out of the old city's rubble (**figure 11.2**), and although some of the wire rope detailing looks a little clunky now, the speed of the design and erection of the structure made it a fitting symbol of the new Germany and its *Wirtshaftswunder*.

11.2
**Olympic stadium, Munich, 1972**
Built over embankments of bomb debris, this cable net consciously expresses the *Wirtshaftswunder*. Earlier experiments with park pavilions in Cologne were extrapolated to very large scale then enclosed with crude laps of sheeting. Sophisticated analysis coupled to simple construction met very tight deadlines.

Some critics cite the pioneers of membrane roof design as the source of our contemporary preoccupation with free-form buildings. At the time their intention was only to create taxonomies of appropriate form. They published comparative studies of tent types. Instead of imagining unbounded shapes, they rigorously and systematically identified and explored the shapes, potential applications and limitations of flexible surfaces.

Otto proposed industrial structures, drop-shaped storage tanks and repetitive shed roofs, as well as more exuberant confections such as exhibition buildings to display products of the new economy. There was no technical differentiation between building types; the variety of forms and their proportions flowed directly from model studies – soap bubbles on wire boundaries and stretch fabric miniatures. Utility would provide its own aesthetic. This open-ended approach to structural design contrasted with less radical trajectories, and other, older conventions of German engineering were resuscitated in the reconstruction. The civil engineer Fritz Leonhardt (1909–1999), a graduate of Stuttgart University, founded his own office just before the outbreak of the Second World War and was well placed to take part in the infrastructure replacement needed after it. Autobahn viaducts and bridges reinforced the substance and activity of West Germany in its contest with the East. The physical geography of the country, a ripple of drainage basins with wide rivers flowing north or south, had led to an inordinate number of crossings being destroyed in the east–west retreats; the strategic need for cross links was replaced by a similar orientation of trade.

Leonhardt and others rushed to rebuild bridges, the earliest replacements being pragmatic and economic designs of which the riveted open girders of the Kennedy bridge in Bonn are typical. Other schemes incorporated nostalgia, direct replacement of familiar outlines. The Hohenzollern bridge in Cologne was rebuilt to a 1910 design of riveted, trussed arches, then widened with matching arches of continuously welded elements.

## The persistence of technological propaganda

Gradually bridge projects began to reflect more modern construction processes; they were regarded as sites to exhibit engineering expertise. Usually set out with classical proportions, they started to show clean, uncluttered lines adopted from production design, particularly that promoted by the influential Ulm School (1955–1968). The Dusseldorf family of bridges, four cable-stayed crossings of the Rhine, includes Leonhardt's Knie bridge, completed in 1969. At 320 metres main span, it was the longest stayed span of its time and embodies the visual balance and simplicity in its harp configuration that its author believed essential. The grouping of bridges at Dusseldorf has become an evocative image for the city.

Leonhardt went on to develop a treatment for the larger autobahn viaducts. Setting up tall, slender pylons, he jetted deep box girders out at high level across the wide divides. On the Neckar valley bridge in Weitingen, visual impact on the landscape was minimised by reducing central pier and deck beam sections and sharing torsions and lateral loads with heavier double columns tucked away below the approaches. The 1982 publication of his book *Bridges: Aesthetics and Design* proved exceptionally influential in propagating Leonhardt's theories of visual composition.[5]

## Focused development

Other German contributions to the practice of bridge engineering included that of Franz Dischinger (1887–1953), a Berlin-trained theoretician. Through studies of the creep and shrinkage of concrete, he gained control of these effects and released new levels of economy in pre-stressed concrete technology. His experimentation was wide-ranging, from very thin shell structures to very thick external tendon systems, and he completed the first externally pre-stressed concrete bridge at Aue in Saxony in 1937. Dischinger's work was often a direct response to commercial pressure, particularly the periodic shortages of high-grade steel in the post-war years, and out of such research came practical improvements in construction methods as well as design. Among his apprentices, Ulrich Finsterwalder (1897–1988) perfected pre-stressed cantilever construction, the sequential erection of bridge sections to meet at mid-span. As the two sides reach out to one another, continual adjustment of the tendons maintains internal stress conditions within manageable limits.

## Crane design

Industrial sponsorship of research had played an important part in engineering development in Germany since the late nineteenth century, and many of the methods of elastic analysis applied to building frames stemmed from theoretical work paid for by crane manufacturers improving their designs. German heavy industry needed ever increasing scales of lifting device and the well-defined structural problems arising appealed to analysts and mathematicians. Sophistication of structural design was a selling point in the detailed brochures expected of mechanical equipment manufacturers, and so new methods of structural analysis were published and promoted.

## Very large transports

Transport links brought economic growth to Europe and strengthened its potential as a trading community as new roads and bridges reached south and east around the Mediterranean basin. The world economy was growing on oil. The Greek shipping magnate Aristotle Onassis (1906–1975) founded a business empire buying up superannuated Liberty ships, and despite their structural flaws kept them carrying cargos to generate the wealth from which a new generation of oil tankers could be created. These vessels would surpass warships in size.

The largest battleship ever built, the Japanese *Yamato*, displaced 65,000 tonnes and aircraft carriers subsequently reached over 75,000 tonnes. The peculiar characteristics of a floating tank, filled with liquid just a little lighter than water, meant that tankers could get considerably bigger than these capital ships. In 1950 a typical tanker weighed 25,000 tonnes. Improvements in welding technology, particularly heat treatments to counter brittle failure and enhance longevity, permitted much larger thin-wall vessels, and by 1963 several vessels had displacements of more than 80,000 tonnes; by 1973 there were many ships over 300,000 tonnes each. Hull lengths of 300 metres stayed very close to the height of the tallest towers of the time. The limitations on scale ceased to be structural but were instead bounded by natural harbour sizes.

Reinforced concrete had become the staple of basic building construction during the war years. Subsequently, designers realised that the material's use still lacked a convincing vocabulary of form, and the ways in which concrete was being handled once again diverged, just as they had done upon its first appearance. This was no longer about the way the material's behaviour was to be understood and analysed but rather what were its appropriate uses, construction methods, forms and details. Once again there appeared a series of seemingly complete but incommensurate systems.

## Reinforced concrete construction developed as a universal building system

The bombardment preceding the Normandy landings left the French port of Le Havre in ruins, and Auguste Perret's long career culminated in the rebuilding of the city. Filtering a deep understanding of reinforced concrete's behaviour through a review of building types, his intention was the creation of a universal system of tectonics, a library of forms and collection of details that hark back to Rondelet's *Encyclopaedia of Construction*. Structure was expressed by exposing a building's frame, emphasised with the pattern of joints created by pre-cast panels. An education in classical proportioning systems and the Baroque organisation of façades as 'field and frame' compositions, came together in a motif of 'framing', only vestigially structural. Perret applied this trope to a whole variety of building types all coordinated within a tightly controlled urban planning system, and the rigour applied to seeing these principles through has made Le Havre a UNESCO World Heritage Site.[6]

With a consensus on the conceptual model of reinforced concrete's behaviour established, it became obvious that development would centre on construction method and delivery systems. Repetition of forms and expressions was initially fostered by the efficiency of using prefabricated and standardised formwork, reusing the lost part of the casting process. This imprint of replication is casting's truest decoration.[7] Pressure to industrialise building, initially for speed and economy and later to improve the safety and comfort of workers, then transformed the multiple use of formwork into the pre-casting of elements. Components made adjacent to site or remotely in factory conditions or covered casting yards achieved better quality than in-situ work. More intricate surface finishes were achievable; failures could be remade and longer curing times allowed. A variety of joints, cast in or mechanically linked, were experimented with. Building units might be sized to suit standard craneage while civil engineering scale works might have special lifting equipment designed to place much larger elements.

## Pre-cast concrete

French-educated engineers contributed much to the art of pre-casting, simultaneously collaborating and competing with the advances made by the German engineers mentioned above. In a relatively conservative profession, the imperatives of bridge-building met with a tradition of innovation in the use of reinforced concrete to attract the technically restless to large span and viaduct design. The Alps were to be opened up, as well as the mountain barriers of central and eastern Europe and the poor, isolated lands of southern Italy. Long stretches of elevated roadway were required to achieve the long curving motorway alignments.

Pre-cast systems for decks and box beams were perfected in the work of specialists such as Jean Muller (1925–), a collaborator of Freysinnet's. These engineers were inventive, driven towards big solutions by the real savings to be made from bold innovation in large-scale civil engineering projects. Muller developed match casting, each segment of a pre-stressed viaduct cast against its preceding neighbour to ensure a perfect fit. His construction techniques were as protean as his fabrication methods. It had become common to erect box-girder bridges by balancing cantilevers outwards from supporting piers to link together. In sensitive landscapes Muller avoided irreparable damage to valley floors by jetting his structures outwards on temporary frames then building piers accessed only from above, the intervening ground remaining undisturbed. The experience gained formed a kind of engineering 'pre-adaptation' for the now common use of jetted construction, building and moving entire bridge decks out into the void, which has culminated in the achievement of France's Millau Viaduct.

## Edouardo Torroja

Engineers from further south seemed more ambitious for a completely new construction aesthetic. Candela has been mentioned. Eduardo Torroja y Miret (1899–1961) was another precocious Spaniard building upon a background of practical understanding. Having inherited his father's contracting firm, he embarked on a programme to develop built form in reinforced concrete, going back to first principles and trying to exploit the material's capability to be cast into plastic shapes. He created a very wide range of structures, including many large arch bridges, combining simplicity in the details with a sure touch in the profiles and weathering surfaces. His buildings are balanced compositions; he experimented with space structures, shells and folded plates that rely for stiffness on their three-dimensional shape. His facility with freehand pencil sketching was mimicked by others and became an accepted way of accessing the possibilities of concrete construction, and his writings record his self-conscious objective to elevate structural engineering into an art, one generating its own context.[8]

## Ricardo Morandi

The Italian contribution was characteristically lyrical. The work of Ricardo Morandi (1902–1989) reflected the distinctive ambience of Italian post-war design. His working

drawings are graphic masterpieces. The Autostrada del Sole was a concrete ribbon intended to cement north and south together in the new industrial age. Morandi's bridges for the project had large classical arches bestriding the valleys. Spandrel piers were elegantly proportioned, details displaying their function and pins and hinges focusing forces. In his buildings, transportation halls and garages, the roof designs began to be shaped to reflect the distribution of forces within, elegant envelopes and folded plates of reinforced concrete.[9] An anthropomorphic poise permeates his work.

Morandi experimented widely with bridge types, with projects in South America providing a context in which his outlook was free enough to condense. Brutal and direct – just a diagram realised – his Maracaibo bridge picks up on the particular design ethos of its time (**figure 11.3**) and is a clear statement of the considerations that bridge design should encompass.

## Pier Luigi Nervi

The Lombard Pier Luigi Nervi (1891–1979) is recognised as the master of ferro-cement construction. He extended reinforced concrete and pre-casting techniques

11.3 *(opposite)*

**General Rafael Urdaneta bridge, Lake Maracaibo, Venezuela, 1958–1962**

The Italian Ricardo Morandi extended his engineering beyond lyrical applications of graphical statics. The Maracaibo crossing comprises carefully proportioned components combined with rigorous logic into a visually robust whole. This was contemporary architectural theory – 'brutalism', or the truthful and direct handling of materials – transferred to structural design.

into the creation of finely reinforced panels of high-strength cement which became biscuit-thin soffit moulds to in-situ cast frames, slabs and shells. In his invention, Nervi was exploiting a combination of circumstances: a cheap, skilled craft force available to make the moulds and set the close-spaced bars; and centuries of experience using cementitious materials and an unparalleled expertise with the plasterer's trowel, coupled with a benign climate to make his shell proposals workable. The Italian contracting industry, beset by the interests of organised crime established in Italian-American post-war redevelopment, was not a conservative environment and offered the innovative a place to operate from.

Early in his career Nervi began experimenting with grid structures and geodesic frames, culminating in the roofs to the aircraft hangars at Orvieto. Dynamited by the retreating Germans in 1942, these structures proved so robust that the roofs came down without breaking. After the war the motif of intersecting ribs, the Gothic sensibility of northern Italy, reasserted itself through the expression of lines of 'isostatic' stress.[10] This was an odd conceptualisation by Nervi. Initially reinforced concrete floor slabs were given curving ribs, then domes were composed from ferro-cement panels combining to form nets of geodesic lines (**figure 11.4**).

11.4

**Small sports-hall, Rome Olympics, 1960**

The variant of concrete pre-casting styled 'ferro-cement' uses finely moulded panels consolidated by a poured topping. Italian engineers produced exquisite soffit modelling. Purporting to be patterns of force, these 'isostatic' lines have little real meaning; they look like channels of energy, gothic ribbing on vaults.

The patterns were justified as being structurally efficient and expressive of underlying load paths, a superficially appealing logic that overlooks the paradox that form influences structural action and in reflecting it must also modify it. The patterns were arbitrary, fantastic and beautiful.

European experiments found their way back into the North American mainstream, where the reception and reshaping of these engineering ideas into something unique to the United States would follow one of a limited number of routes. Typically, individuals or groups of Jewish-American engineers might import an academically developed technique then expand and adjust it to the procurement base available to them. American industry seemed to more rapidly respond to potential than its European counterpart, first realising it and then ransacking it for commercial opportunity, sometimes thickening the finer lines of an idea. This mountain-moving entrepreneurial spirit combined with old European classical engineering in such men as Bertrand Goldberg (1913–1997) and Myron Goldsmith (1918–1996).

## Marina City, Chicago

In the architect Goldberg's hands pre-casting became a compositional device, not so much an expression of manner but rather of repetition building into a larger whole. The paired towers of Marina City built in Chicago's downtown loop area make up a complete residential complex, with flats raised up above an entrance plinth of car-parking spirals and the in-situ cast work of core and ramp carrying ranks of beautiful pre-cast balconies. A single mould is sized to front living spaces and then suppressed subdivisions added as necessary to accommodate smaller sleeping spaces.[11] The serried façade units develop into the famous 'corn on the cob' exterior (**figure 11.5**). On other projects double curved panels are combined and inverted to create large façades rippling within a single field. Learning how to cast, cure and lift such very large panels later proved useful in creating the post-modern confections of the architect Michael Graves (1934–) and others.

The engineer Goldsmith, meanwhile, systematically studied the 'super-tall' building,[12] defining the mega-frame, bundled tube and braced tube concepts that have become commonplaces in skyscraper design. Combining academic research with consultancy for the American modernist architectural practice Skidmore Owings and Merrill (founded 1939), he set the engineering of tall buildings onto a more rationalist basis than that used for the pre-war giants, trabeated structures such as the Empire State Building in New York. Emphasis was placed on lateral bracing systems and the integration of other essentials, vertical circulation, service floors and fire barriers, into the overall structural form. Following on from the Galilean recognition that there is a scale effect in nature – that the weight of an object increases according to its volume while its strength increases as a function of area – Goldsmith showed that the ideal structure for increasing sizes of tower progressed in quantum steps. He conceptualised the notion of the mega-frame;[13] the highest towers were conceived as structural entities with the whole containing the part (**figure 11.6**).

11.5 (*opposite*)
**Marina City apartments, Chicago river, 1959–1964**
One of several essays in pre-cast concrete by the architect Bertrand Goldberg, where construction process overpowers the functional design. Over a lower tier of car park, a helical ramp supports serried ranks of balcony fronts. The standard moulded units ignore aspect and require crude sub-division to meet spaces behind. The silence of the composition has entered the literary consciousness through the descriptions of J.G. Ballard.

Through work with his pupil and close collaborator the Bengali-American engineer Fazlur Kahn (1929–1982) and graduate students of the Illinois Institute of Technology, he began an organised search for optimal form, only now being superseded as the requirements of natural conditioning come to the fore.[14]

## Modern foundation engineering

Huge resources have been wasted down the ages on building foundations – either expensively over-designed, or costly in terms of damage or collapse when undersized. Very tall buildings not only impose massive deadweights on underlying soils but also large fluctuations in stress levels as winds gust and recede. The scale-change in foundation design required huge improvements in assessment and prediction techniques to be made, particularly as the construction of very tall buildings spread worldwide, away from the rock shelves beneath Manhattan and onto the silt deltas and earthquake-riven margins of the Pacific rim and South America. Mass housing and

11.6
**John Hancock Center, Chicago, 1965–1970**
The 'mega-frame' concept is not as obviously economical as it may seem. Forces are concentrated rather than dispersed. The great tower designers such as Leslie Robertson sought instead to utilise every part of the essential structure, beams and columns, to their limit.

infrastructure programmes, in multiplying the requirement for the simplest of components, needed safe, economic footing designs. Soil behaviour, particularly the interaction between solid material, granules or clays, and the groundwater within it, was finally given plausible explanation. Long-term settlement, consolidation as water seeped away from load points and through permeable surroundings, could be predicted and managed. Idealising such a complex medium as ground mass is not easy, and a background of empirical testing informs much of the predictive modelling now in use.

## Karl Terzaghi

The shaping of the science of 'soil mechanics' was very much the work of one man. Karl von Terzaghi (1883–1963) was born in Prague of Austrian descent and studied mechanical engineering in Graz. He became involved in building hydro-electric installations, and from there in assessing geological conditions around dam structures. Surviving active service at the beginning of the First World War, he found a research post at the Royal Ottoman College of Engineering in Istanbul and began to classify and test soils and soil formations systematically. At the war's end, he took his work to America where, despite some frustrating obstructions, he fleshed out a complete 'science of soil behaviour'.[15] In reaction to the nay-sayers, he strove hard to give his work a rational basis, drawing heavily on his training in classical mechanics for this background in theory. A long and successful consultancy gave his approach first place in the groundworks industry.

## Structural engineering as industrial product design

Other immigrants to America were exploring the consequences of industrialisation in life and building. Konrad Wachsmann (1901–1980), born in Frankfurt, trained and then taught carpentry in Berlin and Dresden. As an employee of the largest timber building manufacturer in Europe between the wars, Christoph and Unmarck AG, he designed a standardised and prefabricated housing system. This experience he transferred to all-metal construction in collaboration with the German architect Walter Gropius (1883–1969), architect, teacher and celebrated pioneer of modernist design. At the onset of the Second World War both men found their way to teaching posts in America. Waschmann was to instigate a disciplined exploration of tectonics, concentrating particularly on how small standard components could be composed into a variety of larger wholes, and his research and writings arrived at an almost mystical, rather Teutonic, apotheosis of the joint: the junctures between things and the act of connection itself. Waschmann theorised on infinitely extendable and universally applicable building systems.[16] Working for the US military, he made designs for vast aircraft hangars built up of highly refined components, demountable and transportable, prescient for the new imperialism. Given the costs of precision engineering and the tolerances achievable at the time, his schemes, had they been realised, would have been hopelessly compromised by slippages across the arrays of joints. There is an odd dislocation between the intellectual's theorising situated on these proposals, and the expert craftsman's indifference towards the glaring practical problems.

11.7
**United States
Pavilion, Expo '67,
Montreal.**
High technology
proliferated in the
cold war. The US
pavilion at the 1967
World Exposition in
Montreal comprised
a three-quarters
spherical space
frame twenty
storeys high.
The inventor of the
geodesic dome,
Buckminster
Fuller, backed his
designs for minimal
weight structures
with an obscure
formulation,
structural
'tensegrity'. Claims
to have solved over-
consumption and
environment control
problems reduced
to alternative 'drop
city' applications
and early warning
radomes.

Almost simultaneous with Waschmann, but outside the academic mainstream, Richard Buckminster Fuller (1895–1983) was inventing and experimenting
with industrialised building. Born in Milton, Massachusetts, Fuller later claimed that
his childhood experiences of making and finding in the landscape gave him a
propensity to design and a feeling for materials. His life was a harsh one, and
repeated setbacks gave him an extreme determination to transform the world
through an idiosyncratic contribution to technology.[17] His greatest successes were in
perfecting the geodesic dome and proposing the more general principles of
'tensegrity' structure, a half-resolved structural concept involving multiple load paths
and the 'apparent' stiffness gained by a continuous weave of pre-stressed wires
between struts (**figure 11.7**). From an apprenticeship as a machinist he learnt
sheet metal working, and his proposals for industrialised dwellings, 'Dymaxion'
houses, mix a profound understanding of pressed metal fabrication with homespun
philosophising on man's occupation of the earth.

## The relationship between architect and engineer

American technological development in the immediate post-war years was deeply
influenced by a ready acceptance of the authority that immigrant engineers, designers
and academics were able to exert. In Britain similar power structures directed
developments. The Danish structural engineer Ove Nyquist Arup (1895–1988) forged
close personal bonds with a generation of émigré European architects fleeing the
Nazi persecutions. His life's work was to promote an integration of engineering and
architectural activity,[18] bringing different branches of consulting, structural design,

environmental engineering, transport and infrastructure planning into an integrated organisation. Arup thereby founded one of the most influential engineering firms of modern times, which has branched out into many areas of engineering, leaving the implication that there is a common thread binding the various activities; research and practice were held to be in close relationship. Taking a design from conception to completion with a professional independence was among the corollaries of his approach that have found almost universal acceptance.

## The Sydney Opera House

**11.8**
**Sydney Opera House, 1957–1973**
The occasion of a definitive collaboration between architect and engineer. The physicality of construction was a central concern of designer Jorn Utzon, and structural engineer Ove Arup strove to meet this. The role of structure and its integration became key issues in subsequent building design.

Arup's most famous collaboration was with his Copenhagen-born countryman Jorn Utzon (1918–2008) on the design for the Sydney Opera house in Australia. An international design competition was won with a sketch into which a lot could be projected but which contained little resolution of the building task proposed. Giant shell sails were to be set up on a plinth into which seating banks were excavated, the whole ensemble billowing forward from its site on Bennelong Point towards the harbour entrance. The realisation of the project became a protracted tragedy.[19] The engineer's contribution to events has been portrayed as a measured search for the structural system that could realise the architect's almost unbuildable vision without undue compromise, but the relationship was rather more complicated than that. Utzon had an extremely lyrical take on structure, seeing in it not a Baroque of modern form but rather a seamless coalescence of form and light. His pilgrimage church at Bagsvaerd, Copenhagen (1976), goes beyond structure to reach a tectonic whole employing almost agricultural building components. As the son of a naval architect

he believed he had an intuitive feel for structure, and as Arup strove to realise the preliminary concept, Utzon began major changes guided only by instinct. The expense of the building became notorious, and even with a loss of amenity from the original proposal the Sydney Opera house as built is mawkish, the celebrated pre-cast panel construction of the vaulted roofs a sad shadow of the competition-winning design (**figure 11.8**). Today the dynamic balanced composition of shells no longer seems as unattainable as it once may have appeared. The route taken by the design forced the structural designer to declare the original form unbuildable, then work from a proven construction system to make an acceptable silhouette. Whereas nowadays the techniques needed to deal with the problems posed by the competition entry could almost certainly be conjured up (and were probably readily available then in adjacent industries), at the time the engineer's authority was employed only to say decisively what was buildable and what was not.

## The 'high-tech' movement

It was to be the last time that a technical sanction would be used to dominate a design. Architects subsequently took the idea of structure away from the engineer's hands and made it something they could themselves shape as well as be guided by. One direction was towards a dream of technology and tectonics, the confections of the high-tech style. The other way was to appeal to a new materiality. The style referred to as 'Brutalism' saw a re-appropriation of engineering by architecture, which ebbed back into engineering practice and research.

The acceptance of plastic theory for structural design has been described above. Hunstanton School, by Alison (1928–1993) and Peter (1923–2003) Smithson has been identified as the origin of brutalist design in Britain.[20] Frame and infilling walls are clearly expressed; materials are used with rawness and meet one another directly. The trabeated frame of the building complex was all designed using plastic theory, and the black-painted armature has welded joints and elementary junctions suited to the assumptions of the method. The design approach was perfected with such buildings in mind.

## The Tacoma Narrows bridge

Despite open exchanges following the Second World War, the differences in engineering cultures each side of the Atlantic persisted as highlighted in the responses made to a particularly dramatic disaster. In 1940 the Latvian Leon Moisseiff, mentioned earlier as a key contributor to the engineering of the Golden Gate bridge, designed a very gracious and high suspension bridge over the Tacoma Narrows gorge in New York State; he took deflection theory a step too far. The crossing was an economic design intended to open up an isolated area upstate, but the federal government refused the bill for a more traditional design and so Moisseiff produced a scheme two-thirds the cost which was adopted. On completion, the deck showed a propensity to vibrate in light wind and twelve months later, in a stiff, steady but not extreme gale, shook itself to bits, snaking up and down rhythmically before beginning to twist and finally to thrash out of bounds. The film footage of the event is a familiar teaching tool used to demonstrate how nature can still call forth unforeseen phenomena. In fact, the aero-elastic behaviour of the very slender deck with its flat-sided and extremely flexible deck girder had been cause for concern from inception,[21] and at the time of the collapse wind tunnel tests were under way to find a solution to the problem. Monoplane designers had been encountering the potential severity of interaction between a flexible structure and steady airstream since the First World War. As a surface twists, the airflow over it is modified and loadings consequently change; a cyclical exchange can be set up. 'Classical flutter' was soon well understood by aerodynamicists. Their interpretation was taken as received knowledge and passed directly to bridge designers so they could implement preventative measures in future.

In North America the reaction to the problem was to over-design, hedging against the unknown. For a period of time, bridges there became even heavier and stiffer than before. If Moisseiff's bridge had been designed down to a price and was too 'skinny', in future there would be reserves of strength incorporated into bridge structures. In England, however, the response to the Tacoma Narrows setback was to go towards the problem rather than away from it. There was a search to vitiate the problem, pre-empt it and neutralise it. A bridge could be given a profile that would actually stabilise itself in an airstream so that if disturbed it would tend to return to a steady state. The theory of an aerodynamically profiled deck, effectively an upside-down aerofoil cross section, was given practical application in the design of the First Severn Crossing in 1966, a symbol of the then Labour government's commitment to the regeneration of South Wales and a demonstration of Britain's 'high technology' intended for export (**figure 11.9**). The steel plate box girder was fully welded and traffic

11.9
**Severn bridge, England/Wales, 1961–1966**
The American response to the Tacoma disaster was to throw weight at subsequent designs. In Britain designers sought cheaper alternatives within the aerodynamic domain. A torsionally stiff wing section based on box beam technology was tuned to stability, using wind tunnel tests.

ran on the upper skin. The aerodynamics proved successful, but the other innovations less so. The design is almost too refined in all its parts, its continuous components difficult to repair and now subject to endless maintenance and improvement.

## Technical beguilement

The Trinity test at Alamogordo in New Mexico in 1945, the first detonation of an atom bomb, brought the nuclear era, which, uniquely, will be with us forever.[22] The war was ending anyway, but this event re-centred popular hopes of technology, highlighting the ambivalences. Here was the limitless energy source always dreamed of, yet the ender of worlds and the filthiest pollutant of man's environment ever manufactured. If technical idealism has thus come to be rendered untenable, initially there was a heroic period, just after the atom was first split, when everything again seemed possible.

As the cold war closed in and the nuclear arms race accelerated, there were urgent efforts to widen the uses of nuclear science. This was a period of strange, diverse and atrophied initiatives. The American submarine *Nautilus* was fitted with a reactor to cruise submerged for as long as the crew could stand it, and the boat went beneath the North Pole in a spectacular demonstration of science realising literary invention – Jules Verne's imaginary voyage of his own fictional *Nautilus*. Such vessels were no longer submersible torpedo boats but a different thing, underwater transports. The strategic significance of limitless concealment rapidly increased the size of submarines,[23] leading directly to the cold war missile platforms cruising beneath the Barents Sea. There were all kinds of experiments, and the accidents and suffering of crews were terrible.

A nuclear plane, a modified Boeing Superfortress bomber, was intended to roam the skies always ready for deployment and in addition to its payload carried a platoon of soldiers. Should the aircraft crash, they were to secure a perimeter around the resultant fallout.

The promise of nuclear technology prompted the most hopeful of palliatives in the face of sometimes seemingly insoluble problems of application. The proposal for a nuclear spaceship, Project Orion, was the zenith – or nadir – of all this activity (**figure 11.10**). A catch-all designation for a number of experiments, Orion envisaged a space ship driven by the impulse of atom bombs exploding behind a pusher plate. A cabin was connected by springs to an armoured shield behind it; bombs were ejected one by one backwards and detonated a suitable distance away so that successive blasts drove the vessel along. The resulting rhythmic stuttering would be ironed out into a steady acceleration on the cabin by tuning the spring system to the right frequency. Models of the things worked and a development facility was set up. The large round staff canteen was sized to represent the cabin of the proposed vehicle, a reminder to researchers of the scale of their objective. The collateral damage of ejecting radioactive fallout into the upper atmosphere was coolly computed; so many statistical deaths. At cancellation, the specification of an adequate shield plate still hadn't been resolved. In its heyday the project had been completely believed in. A linking sequence of the 'technically accurate' science fiction film *2001: A Space Odyssey* was originally to have

11.10
**Project Orion
nuclear spaceship,
1957–1964**
The film *2001: A
Space Odyssey*
originally was to
include a scene
with a large, far-off
space probe
crossing the void on
a stuttering wake of
fire. Meticulously
researched, the
story referred to a
craft then under
development.
Shielded by a
sprung pusher
plate, it was to be
driven along by a
series of controlled
nuclear explosions.
The structure could
be almost any
weight to absorb
the shocks
involved. The
confidence
engendered by the
runaway success of
the nuclear
programme made
any dream seem
attainable.

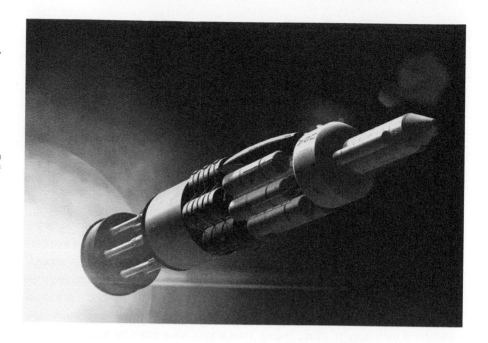

shown such a craft crossing space.[24] That conditions existed between 1958 and 1963[25] for such an enterprise to flourish needs anthropological explanation.

## Aerospace structures

Rocketry developed in a different way. Germany had sought the secret weapon that would transform human conflict and the North Americans found it using European science. Wernher von Braun (1912–1977), born in Poland to German parents, became obsessed by space travel at an early age. It was the central objective of his career, and in the Second World War he transformed rocket design from an amateur pastime of oversized fireworks into a reliable delivery system for bombardment. The V2 rocket that appeared in 1944, built with slave labour,[26] was a true ballistic missile, astonishingly prescient of future developments and only limited in range by the scale that the German economy could manage at the time. It was not a war winner. Von Braun escaped a war crimes trial to become a director of NASA and a founder of the United States Space Program.

The Manhattan Project in itself, the vast organisation and concentration of technical effort required to realise an atomic weapon, was an unprecedented effort. Management techniques pioneered to coordinate the diversity of research, systems theory and mathematical programming that was required became industry staples. Their widespread adoption has shaped subsequent scientific, industrial and commercial enterprises.

Theoretical physics predicted the almost incomprehensibly more powerful thermonuclear weapons that an atom bomb could act as trigger for, and the American military industrial base moved on to their creation. Von Braun's liquid-fuelled rockets were scaled up to carry these bombs into sub-orbital space to land upon and obliterate other continents.

The inertial loads to which space vehicles are subject are accurately predictable, and the design of missile structures, cylindrical tanks and airplane fuselages provided a clear and well-resourced pathway along which structural analysis could be refined. The shape of modern computer analysis began to condense at this time around these tasks. British wartime decoding successes had relied heavily on the work of Cambridge mathematicians and one, Alan Turing (1912–1954), worked out a theoretical foundation from which to resolve the early hindrances stifling the use of the electric programmable computer. Applications of the calculating power released were rapid and widespread.

In structural engineering, analytical techniques were initially improved by the recasting of existing formulations into matrix mathematical notation, arrays of numbers describing the state of things and a set of operations to transform those states.[27] Previously the structural analysis of complicated forms had revolved around searches for simplifying assumptions and then the application of numerical methods. Armies of so-called 'stress-men' were employed by the aeronautical industry to make the multiple calculations necessary to elastic theory. Their efforts were aided by adding machines brought across from the business base transforming itself in the United States, and a predilection for computing and computer-based methods was established.

## The piecemeal that is modern structural engineering

The aerospace industry generated a range of structures with shapes and construction details that could not be compromised to help analysts. These forms were beyond classical solutions. Approximate methods had to be relied on. The development of these techniques was piecemeal, a stitching together of several different discourses. If the monster that resulted still shows its joints, it nevertheless asserts its universal applicability and threatens the destruction of all alternatives. Modern structural analysis bears the traces of at least three conceptual bases.

First are its mathematical foundations.[28] In 1953 an intuitive commonplace amongst mathematicians was definitively proven by a Russian researcher. L.I. Kamynin showed that the method of finite differences, invented long before by the English mathematician and Royal Society alumnus Brook Taylor (1685–1731), was reliable. Small finite quantities, amenable to computer manipulation, replace the infinitesimal quotients of partial differential equations, mathematical models that describe physical processes, which would otherwise be insoluble. This intellectual tool took discrete points along a continuum of some kind, a mathematical function or a curving structural element, then found the conditions at those points and in between.

Following on from this development came another independently derived but complementary analysis tool, the method of finite elements. Richard Courant (1888–1972), a German-Jewish mathematician, had come to America in 1933. Ten years later he found some particular approximate solutions for a series of vibrating systems and recognised that he had created a variation on the so-called 'Raleigh-Ritz' method of numerical analysis. This provided the foundation for a more general method of breaking a continuum, surface or solid, into elements – points along the boundaries of which could be used to determine internal conditions. The method was given its definitive form in a paper published in 1956 by James Turner et al. and the name was finally coined by Ray Clough in 1960. Not until 1973 was the 'finite element' method formally shown by Kadir Aziz to be universally applicable to problems of functional analysis. Formally insoluble problems were now completely tractable.

## Tools framing problems

The finite element method developed as a theoretical concept discussed among mathematicians before the advent of the electronic computer. Such numerical tools shaped the way machine calculation developed into the handling of large streams of data. At the same time structural analysis problems had to be framed in ways amenable to the business-based computers which the market called forth, and the influence of these platforms is the second definitive part of modern structural analysis. Other computational architectures, perhaps more suited to the handling of built form, fell by the wayside. The third component of contemporary structural theory is the choice of idealisation. The notion of analysing a continuous structure by imagining it composed of small elastic bars, the lattice analogy, dates back to the suggestion of the German physicist and engineer Karl Wieghardt in 1906. This substitution of an imaginary structure at some remove from reality proved so powerful as to overrule objection and is now ubiquitous. When Egypt's first Aswan dam was raised in the early 1930s, its cross section was imagined as a fine lattice and the internal stresses computed by an early application of finite element theory involving thousands of repetitive calculations using an aircraft company's design office. So when the need arose for an all-powerful structural design aid to enable the design of airframes and aerospace structures, a theory was available which matched with early approaches to three-dimensional form and which also happened to suit computerisation. This separation of development strands has meant that our contemporary tools of analysis are of little or no consequence in the generation of built form. At the same time that the innocence of technological innovation was being lost, so too the arrival of universal method was signalling the end of engineering.

Other mathematical tools developed at this time have proved only slightly less significant. Experimental stress analysis, not only prototyping but also the monitoring of structures in service, was made practical by a widespread application of statistical tools and data logging. One of the oldest of engineering techniques, the simple act of observation, sidelined by the complexity and apparent inscrutability of modern building, has been given a new lease of life by computing programmes that assemble and assess data. Well controlled, this information yields rich insights. Statistical and set-theory mathematics have proved capable of dealing rationally with uncertain conditions and have begun to reshape the basis on which conditions can be anticipated. Increasing accuracy in determining load spectra will be the biggest single advance in structural economy in the coming century.

Chapter 12

12.0 *(opposite)*
**Bavinger House,
Norman,
Oklahoma,
1950–1955**
An example of a
very light form of
engineering. A
series of north
American houses
use *objets trouvés*,
often junk from the
war, Quonsett
frames and
plexiglas bubbles,
with an intimate
understanding of
minimal structure
gained in the
automotive and
aircraft industries.
The Midwestern
architect Bruce Goff
and his engineering
collaborator
J. Palmer Boggs
raised these
collages pitch-
perfect on tightly
controlled
geometries. Bruce
Alonzo Goff,
American,
1904–1982,
Bavinger, Gene and
Nancy, House:
Design and
Presentation
Drawings, 1950–51,
graphite pencil on
tracing paper,
variable.
  Gift of
Shin'enkan, Inc.,
1990.811.1–28(.14)
elevation opposite
entrance (west
elevation), approx.
60.5 × 81.5 cm,
The Art Institute
of Chicago.

# The continual present (1950–2000)

It is impossible to observe one's own time objectively, to see the wood for the trees, and perhaps immoral to attempt any assessment of it. But how can one see another time more clearly? Speculation is inevitable, and this section on contemporary work might perhaps be best considered fiction, like Wells's *The Shape of Things to Come*.[1] The presentation will be partisan. Given the sheer diversity of current viewpoints, there's little gain in adopting Delphic technique in the hope of catching the future in some kind of consensus, either overt or subconscious. Instead, by close analysis of examples we'll detect in the continual present some of the structures and tropes within the work of current practitioners and examine them done well and their combination and enrichment.

The past fifty years, since the economies of the world began properly to recover from the Second World War, have seen the development of two main strands of structural design engineering. On one hand the protagonists themselves have defined an architecturally orientated engineering, and on the other there comes a resurrection of engineering's potential as an art in itself.

## Collaborations between architects and engineers

The relationship between architect and engineer has returned as the subject of self-conscious speculation and experimentation. The absence of consensus and a general instability and diversity of output needs explanation. There is a certain periodicity about this: Victorian debate then definition, productive consensus, dislocation and debate and so on. A number of pairings, colleagues working together over extended periods, demonstrate how architectural enquiries and engineering approaches of individuals adjust around one another.

J. Palmer Boggs typifies the engineer working at very refined scale, immersed in detail, very closely following an architectural lead. He supported the work of Bruce Goff (1904–1982), who produced extraordinary buildings in the Midwest of North America, mostly from lightweight materials and industrial high-tech detritus. Both men had experienced war serving with the US Navy Construction

Corps in the Pacific: America moving westward with a contingent infrastructure of temporary dwellings and service buildings. The best standards of accommodation were sought together with rapid deployment.

The Quonset system of curved metal formers carrying profiled sheeting made huts and hospitals. The form can be traced back to the council lodges of the Iroquois Indians.[2] The distinctive characteristics of such assemblies were their ease of manufacture, suitability for transportation and simplicity of assembly. The joints between the skin and ribs of the Quonset hut, all steel, were made with nails driven through the sheets to pinch between a partially welded seam along each former. Goff and his engineer experimented with this attitude, taking components and materials, plastics, even anthracite and bringing them together in new patterns.

The Seebee Chapel at Camp Park, built in the last year of the War, transforms an industrial building system into middle-brow art. Standard ribs were lifted out of context, blacksmithed into spirals and suspended off derricks or tripods. This strand of contingent engineering and collage architecture has re-emerged most recently in the guise of buildings largely composed of subcontractor design elements, encouraged by shrinking consultancy fees and shifting competencies in technical design. Working as an architect, Palmer Boggs's interests shaded into early examples of low-energy and environmentally sustainable design.

August Komendant (1906–1992), engineer and architect, collaborated with the great theorist and influential architectural thinker Louis Kahn (1901–1974), taking the ideas of Buckminster Fuller (see page 178) into mainstream architecture. An ideal of a modern architecture deeply informed by structural principles and servicing strategy found expression in a series of their projects.

The Richards Medical Research Library is a demonstration of their earlier theorising. Pre-cast concrete floor units, part cantilever, part grillage, are pulled together with tendons to form deep serviced floors balanced on masonry shafts containing risers. Structure supports Kahn's differentiation of served and servant spaces.[3] Further projects saw a stronger expression of structure as armature. Buckminster Fuller's ideas for geodesic forms and minimal weight frames were tried out at a grand scale. References to classical forms appeared: long barrel vaults suffusing light from ridge openings, brick arches of imperial scale, either tied the way openings were once reinforced in the rubble walls of later antiquity or reflecting themselves into relieving arches like the Pons Fabricius (Chapter 2, pages 34 and 35).

Both Palmer Boggs and Komendant maintained careers as architect as well as engineers and the re-combination of roles in individuals presented diversely. One of the most famous architects of his day, Frank Lloyd Wright (1867–1959) had started out in a civil engineering department. Throughout a very long career Wright's work environment was saturated with entrepreneurial engineers modernising the Midwest. He apprenticed with the partnership of Adler and Sullivan, engineer and architect respectively, collaborating on the earlier skyscrapers. Occasionally clients were engineers or industrial engineers. This propinquity appears to have encouraged his continuous efforts to justify his designs by structural logic, leading him to some bizarre speculations on the physical behaviour of buildings. Many of his projects stand up not by any physical resolution but by not being at too large a scale. Despite his rationalisations and theatrical load tests, the lily pad hall of the Johnson Wax Administration building works by convoluted load paths. The cantilevered balconies of 'Falling Water', one of the most celebrated private houses ever built, have never worked. The raft foundation of the Imperial Hotel was presented as the saviour of the building in the Great Kanto earthquake of 1923.[4] In fact the building was sited in an area unaffected by the ground shocks. Wright went through a number of collaborations with engineers.

## Lightweight steel houses

One post-war programme for new housing in North America made particular use of structure to inform design. The 'Case Study' houses were a series of projects for self-contained houses in the West Coast landscape, even then an American dream of questionable sustainability. Between 1940 and 1960 John Entenza (1928–2006), publisher and editor of the magazine *Art and Architecture*, instigated and promoted these experiments, promoting the steel-framed house as a type with a range well beyond its obvious limitations. Lightweight frames, tubular columns, simple beams or hairpin trusses were used as the controlling grid for lightweight panel walls and glazed screen envelopes, dissolving divisions of inside and out. The industrial designers Charles (1907–1978) and Ray (1912–1988) Eames, created a house and studio out of abandoned components found on site, a portalised armature and collage of panels, completely rational yet a tightly controlled composition relating to the surrounding stand of trees.[5]

Another contributor to the programme, Craig Ellwood (1922–1992), started out as an estimator before going to night school to study metal construction. His simple planning rapidly improved to match a sophisticated use of brake-pressed steel components, plastic siding and glazing units. Clearly influenced by the formal exercises in house planning made by Ludwig Mies van der Rohe (1886–1969), he close-coupled making with form, totally at odds with Mie's indifference to practicality.[6] A forgiving climate and confident demeanour allowed this Hollywood architect to keep up a stream of experimentation in construction, from junction details to industrialised systems. His use of tilt-slab technology, raising walls cast on the ground, like barn building, to make factories, reflects his self-imposed remit always to find appropriate expression of the new ways of building.

Two architects, Raphael Soriano (1904–1988) and Pierre Konig (1926–2004), produced the best versions of the steel-framed house development. Both concentrated on the framing 'in itself' before moving on to incorporate it within the larger whole. Soriano made modular prefabricated systems in steel, aluminium and plywood. By limiting himself to the barest of means, a single column size, a single beam depth and a simple profiled roof decking, Konig came up with one of the most influential and persistent images of the modern house. The Bailey House, Case Study house No. 21, is the simplest of L-shaped plans, a bedroom wing embracing a pool,

12.1
**Bailey House, Case Study House 21, Los Angeles, 1956–1958**
A certain elegance is accessible through the rigorous pursuit of a minimum of means. Pierre Koenig's steel house in the Los Angeles hills has become a cinematic staple. The frame is reduced to a pair of rails on circular posts. Deep trough decking lays directly over. Junctions and sealing systems are direct and simple. The temperate climate helps. J. Paul Getty Trust. Used with permission. Julius Shulman Photography Archive, Research Library at the Getty Research Institute (2400.R.10).

living accommodation cantilevering out from the hillside to look down on the Los Angeles conurbation. Julius Shulman's (1910–) photograph of the glass eyrie above the street lamps of the city's gridiron initiated a genre (**figure 12.1**).[7]

## Cold war design

The onset of the cold war and the loss of colonial empires made huge impacts on Europe. England particularly reacted to a loss of prestige by promoting an appearance of extreme technical sophistication. Complicated infrastructure and bridge designs appealed to the export market and helped maintain presence across the Commonwealth. The British aerospace industry felt it should achieve supersonic passenger transport. The cultural component of the struggle between the superpowers manifested itself in a particular style of consumer design suitable for the advertising industry to illustrate. A drive to express structure as an intrinsic part of modern design, in product as well as building, brimmed over into 'high-tech' style. The phrase traces back to an article of 1957 linking the dense populations of Western Europe and their background of highly developed technologies to the requirement for nuclear power. In construction, it manifested itself as a mannered emphasis on modern materials, their junctions and crossover technologies. Energy use was ignored. Manufacturing systems from the automotive and aerospace industries were jump-cut into the stream of construction improvement, still most primitive, with little time for adjustment or assessment. This phase produced an iconic engineering, structural expression exceeding all functionality.

## Frank Newby

In the Americas and all around the Pacific Rim, the leading edge of construction innovation concentrated on super tall buildings and problems of constructability. The post-war austerity in Europe and Britain persisted as an influence, with emphasis placed on smaller, exquisitely detailed projects. The type of structural engineer who not only works within the logic of his craft but also attempts to assimilate the architect's enquiry and to adjust his or her way of working to that frame of reference was first fully embodied in Frank Newby (1926–2001), partner of the firm Felix Samuely and Partners. Samuely's was one of a number of London structural engineering firms founded by émigré engineers (Stefan Zinn, Jan Bobrowski and Wilem Frischmann are also important names), just before the war who fearlessly promoted new methods in the professional voids opened up by the conflict. From this fertile British field, Newby became the first of a generation of structural engineers who were to go 'bush' in their design sensibilities. This odd English euphemism best describes the way architectural understanding was embraced. In the process of assimilation, an outsider will firstly display then eventually over-emphasise indigenous traits.

Newby took up the enquiries of those around him with gusto, producing the extraordinary 'Skylon', a structural folly, at the Festival of Britain (1951), in collaboration with the architect Hidalgo ('Jacko') Moya (1920–1994) (**figure 12.2**). The young pre-stressing expert jacked up the sagging tendons of the constructivist confection each and every day of the six-month exhibition.[8]

12.2
**Skylon, Festival of Britain, 1951**
Pre-figuring the space race, the Festival of Britain included an elemental structural sculpture to express the aspirations required of the country. The cigar-shaped frame relied on integral lighting combining with a riverside siting for much of its impact. The young project engineer, Frank Newby, would re-tension the cable system each morning, learning the practical shortcomings attendant on a seemingly simple structural diagram.

London's dominance of British building design at the time meant that Newby was not alone and collaborators, competitors and apprentices began to coalesce into a critical mass of engineering ideology. Together with Sven Rindl (1922–2007), an engineer with a sculptor's sensibility, Newby ranged across the borders of engineering. The Snowdon Aviary, with the celebrated design theorist Cedric Price (1934–2002), relied on the cable expertise gained on Skylon. James Stirling's (1926–1992) Leicester University Engineering building was composed with a quite balletic poise. Eero Sarrinen's (1910–1961) American Embassy building on Grosvenor Square combined a favourite structural trope of Newby's, the 'star beam', with large scale structural pre-casting. Kahn and Komendant's influence (page 188), was nuanced with an integrated servicing strategy, the structure evenly permeated with voids instead of cut by jarring zonings.

The playground of British structural experimentation extended into the commercial sector. Richard Seifert (1910–2001) and Wilem Frischmann (1931–) completed Centre Point, sited on London's east–west divide. The thirty-two-storey tower remains a quite extraordinary structure. An exterior envelope of elements, transoms and mullions integrated into deep embrasures, supports column-free floors.

## The collapse of Ronan Point

The industrialised ethos and economy of pre-cast structural concrete promised much. Drawings of neatly interlocking systems in the process of making delighted architects and socialist politicians. But the overzealous promotion of such 'modern methods of construction' brought a rapid hubris. Any number of sodden, condensation-racked, deck-access apartment blocks were thrown up, condemning a generation to instantly substandard housing. In the early morning of 16 May 1968, a kitchen gas explosion collapsed the entire corner of a twenty-three-storey block of system-built flats, Ronan Point, social housing in the London Borough of Newham. The pack-of-cards failure, a progressive collapse initiated by the loss of a single panel, struck straight at the core of public and personal confidence. That such a modern product could be quite so simply unsafe and that social provisions had hurt people were a wake-up call to engineers and their legislators.

## Anthony Hunt

Often it is the second generation that is most fruitful on prepared ground. Out of the Samuely milieu emerged Tony Hunt (1932–). Educated at London's Brixton School of Building, he was above all a realiser of ideas. He recognised that structural engineering embodied a design function but also that in itself it was degenerating in the high-tech cul-de-sac. Engineering requires a meta-system of principles to guide it, an enquiry to shape its approach. His early contribution, working with a youthful firm of architects, Team 4,[9] came within that English sensibility which synthesises the initiatives of others. Together the group first tamed 'Japanese metabolism', a theory of infinitely extendable and re-configurable architecture,[10] by applying it to the problem of high-specification manufacturing facilities and office parks. A project for Reliance Controls

12.3
**Inmos factory,
Newport, South
Wales, 1980–1982**
The 'high-tech'
architectural style
was well served by
British engineers.
Function and
construction were
exuberantly
expressed to the
point of parody.
Pin joints cleanly
articulate
components.
A superclean
environment for
the assembly
of computer
components, this
factory implied an
infinitely extendable
space radiating
from a service
access. The outer
props were a late
addition to the
design.

(1967) worked within the straight trabeated idiom of rolled steel framing and profiled sheeting of the Case Study Houses. The Sainsbury Centre, an art gallery for the University of East Anglia (architect Norman Foster and Partners, 1977), demonstrated the idea of an extendable 'cool tube' of neutral space, within a deep envelope and service void which could be reconfigured from without the enclosed space. The clean facility of the Inmos Factory in South Wales (architect Richard Rodgers Partnership, 1980) brought structure and services completely outside, an early indication that building servicing strategies would become the predominant determinant of design (**figure 12.3**). As such highly differentiated designs were reaching extremes, Hunt's boundless verve turned to subtler structural devices in support of architectural expression. All detail is suppressed in the frameless glazing of curtain wall encircling the Willis Faber Dumas headquarters building in Ipswich (architect Norman Foster and Partners, 1975). The train shed for the Channel Tunnel terminal at London's Waterloo Station (architect Nicholas Grimshaw and Partners, 1993), presaged a revived interest in Victorian achievements as a source of design ideas.

Throughout the 1960s in London, the smaller companies of generalists were making the running, perhaps slightly less constrained by market requirements and the received knowledge of the engineering establishment than their bigger competitors. The large British consultancy Ove Arup and Partners responded to the sea change by promoting a star engineer, the Irishman Peter Rice (1935–1991), who perfected an inspirational and straightforward engineering style. A series of experiments in unusual structural materials, ferro-cement, glass and plastic, resulted in exemplary buildings

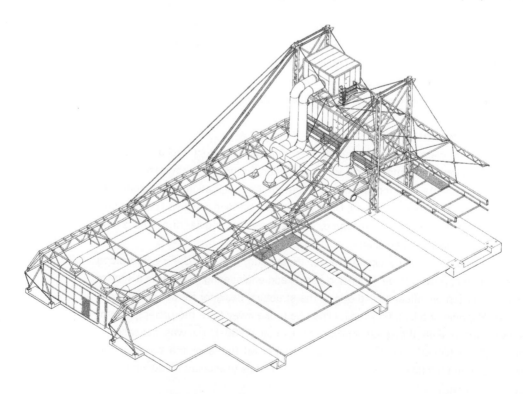

12.4
**Menil Collection
Museum,
Houston, Texas,
1984–1986**
A continuing
commitment to
materials
development
characterises one
strand of structural
innovation.
Composite forms
generate interesting
interfaces. This
Texan art gallery
seeks to modulate
light through a
carefully shaped
baffle system.
Ferro-cement
louvres become the
lower booms of
trusses triangulated
with cast steel
struts. Joints
expand like
knuckles to transmit
loads.

some of which nowadays look somewhat laboured. His approach of careful implementation and working from the intrinsic properties of materials was readily transferable to a subsequent generation. Rice's most successful buildings, attenuated and detailed with a light touch, were made in close collaboration with Renzo Piano (1937–), an Italian engineer who also made a series of material and assembly experiments using the building firm he had inherited, and who crossed the architecture and engineering divide to become a fully fledged building designer (**figure 12.4**).

## Centre Pompidou, Paris

Rice and Piano were in the team with English/Italian architect Richard Rodgers (1933–) that won the Centre Pompidou competition for an arts showcase in Paris (1971). Their scheme was for the long over-determined realisation of the concepts of the teacher and architectural guru, Cedric Price. The building dissolved into a representation of a reconfigurable machine, and expressive engineering was a key component of this semantic shift.

Like the Sydney Opera House, 'Beaubourg', as the Centre Pompidou became known, proved to be seminal in design and engineering. It became a site for much discussion on influence and origin. The structural designer for the project was Edmund Happold, working for Arups. When he broke away from the project to form his own group with the great analyst Ian Liddell (1938–), he was cat-called a 'lightweight engineer'. The slender, articulated braced frame of the building was deemed to need the development of new analytical tools to account for second-order effects, despite the intention that it should be no more than an industrial-grade armature.[11] Ted Happold made a great contribution to the development of design education in Britain through the agency of a combined course for architects and engineers at the University of Bath. The company of Buro Happold made a series of innovative structural forms by appropriating and adapting computer tools from other fields before settling into a pattern of high-quality, integrated commercial engineering, advancing with considered improvements in materials use and detailing development.

## Integrated design

Market forces led Ove Arup and Partners to diversify into the various branches of engineering, and an effort was made to re-integrate the specialisations. It was taken as read that the various engineering discourses have a common foundation and that design itself is single identifiable activity. (The cultural differences within even single sectors of creative industries are only just now being investigated.) A single practice combining all the professions associated with building design was established by Ove Arup and brought on by the Rhodesian Sir Philip Dowson (1924–). Teams contributed on a round-table basis to produce buildings equally influenced in their form by all the disciplines. Technology was used very directly. Dowson had been in the submarine service. Initially there were a series of factories, based on metabolist principles, with structure set on tartan grid or serviced aedicules, and materials brought together in a brutalist way. Then, perhaps oddly, followed Oxbridge colleges and corporate headquarters using pre-casting systems and pitched roof in neoclassical arrangements, pavilions and courtyards, all shaped by closely integrated and immutable service provisions.

## Structures for the North Sea

The giant Forties oil field was found beneath the North Sea in 1970. Its exploitation was on the very limit of technical feasibility, and ten years went by before oil prices made the effort economically viable. Floating islands, platforms, static structures of unprecedented scale, were suddenly being installed into one of the world's harshest load environments, open sea and storms. Frames and accommodation modules were procured at the scale of civil engineering works. A generation of engineers gained confidence in handling new magnitudes of forces. Fracture mechanics advanced to address the pounding of wind and waves. Reliability methods, statistical approaches to assessing the likelihood of failure and therefore the direct costs of safety were transferred in from the aerospace industry. As the North Sea work rapidly peaked and began to peter out, the expertise became available to other infrastructure projects and large-scale buildings. The extremes of construction that had been created proved that almost any form was realisable. As descriptive systems, computer-modelling tools, first recorded then started creating three- or more dimensional form, so now structure as a determinant of that form recedes then vanishes. The French Architect Jean Nouvel (1945–) summed up the process: 'I used to pay attention to my engineers. Now they have become magicians and I ignore them. They can achieve whatever I ask of them' (**figure 12.5**).

12.5a and b
**Suspended screen, Kimmel Centre for the Performing Arts, Philadelphia, 1998–2001**
A sophisticated structure: conceptually unlikely, yet simple to analyse. A frameless meniscus of glass is suspended across the end of a barrel vault and resists wind pressures and suctions by bowing against the resistance of hanging ingots set like curtain weights. The structural scheme relies on the long-term flexibility and adhesion of silicon seals between glass panes.

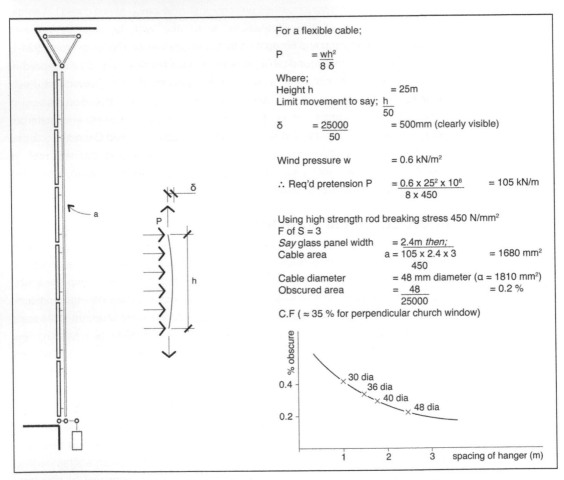

For a flexible cable;

$$P = \frac{wh^2}{8\delta}$$

Where;
Height h $= 25m$
Limit movement to say; $\frac{h}{50}$

$\delta = \frac{25000}{50}$ $= 500mm$ (clearly visible)

Wind pressure w $= 0.6$ kN/m²

$\therefore$ Req'd pretension P $= \frac{0.6 \times 25^2 \times 10^6}{8 \times 450}$ $= 105$ kN/m

Using high strength rod breaking stress 450 N/mm²
F of S = 3
*Say* glass panel width $= 2.4m$ *then;*
Cable area $a = \frac{105 \times 2.4 \times 3}{450}$ $= 1680$ mm²
Cable diameter $= 48$ mm diameter ($a = 1810$ mm²)
Obscured area $= \frac{48}{25000}$ $= 0.2$ %

C.F ( $\approx 35$ % for perpendicular church window)

## Dis-ordered systems

This enervation has prompted a variety of responses. Today's leading structural designer at Ove Arup and Partners, Cecil Balmond (1943–), has perceived order behind the chaos. Modern mathematics dealing with the unpredictable, catastrophic events, and the more mystical aspects of number theory are summoned up in a search for a new definition of form, a harmony in disorder.[12] This work is almost full circle to the ancient Greek idea that immutable pure number is the route we have been given to the true nature of things. Balmond claims to have made a number of definitive buildings encoding transitions to new typologies. Given the complexity of the task that is design and the way human designers have responded methodically, consciously, unconsciously, intuitively and irrationally over the millennia, it may be a little way further before the rationalisation of the miracle of design is fully achieved mathematically or mechanically.

## 'The Art of Engineering'

In north-western Europe the idea of engineering as a province of art persists; the engineer should be capable of a multivalent response, respecting the complexity of the problem in hand. In France, Mark Mimram (1955–) is a designer of expressive structure. Tapping into the Parisian tradition of filigree metal and glass, arcades and verrières, he has produced a series of intensely studied and mannered assemblages, every detail displaying the thought that has gone into it, each component reflecting the way it was made and the properties of its constituent material. An English counterpart, Chris Wise (1961–), formerly a design director at Ove Arup and Partners, has made the intuitive sketch, the esquisse embodying the essence, into the centrepiece of a process of creating structural form.

## Oskar Graf

Putting the sketch centre stage in the design process is a perennial theme, currently ascendant. It should be loose and free of constraint, somehow capturing, without confining, the fleeting moment of idea, the point of anamnesis when the divine spark crosses over. In the hands of an experienced engineer, it gains (presumably) from a deep fluency and familiarity with structural logic. In order to achieve its full potential, some believe, it must come from further within the author, from the subconscious, from the kinaesthetic sense of the hand that made it. It becomes important what medium it is made in, soft pencil, stylus in etching ground, burin on steel plate. The Austrian structural engineer Oskar Graf describes how a lifetime's experience encountering one's own built design guides the spatial movement of the pencil towards a timeless continuity.

Graf engineered several projects of the architectural partnership Coop Himmelblau, who claimed to have adopted an old surrealist device. Sitting in a darkened room they would, on the very cusp of falling sleeping, dash off a sketch, almost like automatic writing, which would then remain central to the design work. The engineer's role was to abstract from the craze of lines a viable, maybe even efficient, structure.

## Santiago Calatrava

The Catalan structural engineer turned architect Santiago Calatrava (1951–) was educated at the Arts and Crafts School, Valencia, and then in the ultra-rational atmosphere of Zurich's ETH (see page 131). He has revived the interpretation of graphical static images as honest structure and amalgamated it with notions of bio-morphology, birds on the wing, growth spirals, to produce a string of white bridges and cantilevered canopies of photogenic exuberance (**figure 12.6**). A daily regime of life drawing in the morning and engineering in the afternoon has enabled him to reconnect structural design with the close study of natural forms. A series of sculptural experiments, the balances, levers, twisting and locking blocks of George Bridgeman's[13] life study system have been reproduced as full-size buildings. In popularising these platonic rationalisations of form, matter and nature, Calatrava has transformed contemporary attitudes to engineering, away from the reductive and back towards a cultural register.

## The proliferation of alternative discourses

As new discourses emerge, older ones review and consolidate their philosophical foundations to stay relevant. Their advocates either hone down an idea for its kernel of truth or embrace and embellish variants, seeking the insights such process can provide. The founders of the building design partnership Rice, Francis, Ritchie were vigorous proponents of the high-tech experiment. No hint of cloying expression detracts from a sequence of building gadgetry, details releasing the architectural potential of modern materials: glass, silicones and steels, and all based on a close reading of manufacturing means and material properties. Yet their method was not automatic but a carefully studied artefact in its own right. The self-consciousness of the designers towards the tradition in which they were working is revealed in an early video of the three sketching on restaurant napkins. They are envisaging a frameless

12.6
**Stadelhofen Railway Station, Zurich, 1983–1984**
The differential equations describing bird flight, growth and stress flows relate and admit the possibility of resonances of form. An almost anthropomorphic balance intrinsic to the stylings of the Catalan Santiago Calatrava is probably more relevant to their immediate appeal.

glass box exhibition building in the same way that Joseph Paxton (see page 114) conceived of his Crystal Palace on a dining car napkin.[14]

The partnership was to split, with the firm continuing as structural designers. Martin Francis turned to boat design, perhaps the best place to achieve beauty purely through technology. Peter Rice's contribution is mentioned above. Ian Ritchie (1947–) kept up a 'through line', running the rational and lyrical as dependent variables in his projects. Rigorous use of materials, a kind of tectonic morality, is inverted from a grammar of construction, limiting form, to being instead a route to the widest range of possibilities. By drawing and communicating in various media, pencil, charcoal, etchings, more nuanced perceptions of form and space are released than those obtainable by computation.

The Dublin Monument to the Republican Uprising of 1916 is profound in its understatement, a 200-metre-high spire of shot-peened stainless steel. The extreme attenuation exploits the latest in analytical method and wind engineering. Internal mass tuned dampers smooth away vibrations from buffeting. Access and lighting maintenance is cleverly considered and the construction method made safe and simple. The object touches the ground with great subtlety. And yet through all this attention to detail, strenuous confrontation of construction problems, it is how the splinter of grey metal acts to animate the city sky, fracturing the light, that is never lost sight of (**figure 12.7**). The surface finish is precisely engineered to its purpose.

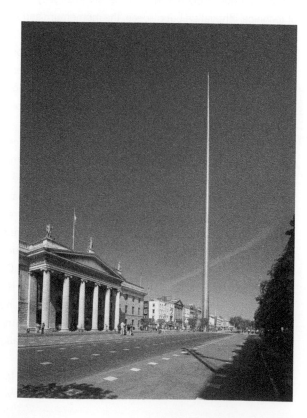

12.7
**Monument of Light, Dublin, 2002–2003**
Modern structural theory's instabilities, aerodynamics and fatigue phenomena are focused single-mindedly onto the achievement of a minimal wand, a monument to republican aspiration. Conventional flange joints, well understood, are chosen for attenuation. Dynamics are controlled by dampers made adjustable to retrieve modelling errors and false assumptions.

## The critical analysis of engineering

A self-critical approach also underpins another developing tradition in structural engineering within which this book probably wishes to find itself. The historian Rowland Mainstone (1923–) has led a stampede of English academics and retired engineers analysing historic examples of built form using the method of technical determinism. His study of Hagia Sophia is outstanding, unravelling the sequence of building programmes and back-calculating the original designer's responses to failures of intuition as they surfaced in crack patterns and bowed masonry. This approach to design, almost as if it were a 'management' of natural processes, finds contemporary application in Jacques Heyman's work as surveyor among the great English Cathedrals. Even better than Mainstone is John Rapley's report on the significance of Stevenson's Menai bridge, contextualising the extraordinary conceptual and economic spaces of the Victorian engineer. A plausible milieu and an intellectual outlook are conjured up, in which the decisions made can be re-connected to the ideas and intellectual tools that shaped them at the time.

## David Billington

There are the beginnings of a real critical framework for engineering forming within this work. The books of David Billington show how the work of individuals such as Maillart and Menn is not just the product of the immediate influences surrounding them but also part of contiguous phenomena.[15] The 'Swiss Legacy' he identifies refers to a history not of technical innovation but of ideas, the momentum and arbitrary direction that drives and directs development. Are there timeless absolutes that underlie all specific incidents of engineering? The naval architect John Coates, contributor to an icon of 1960s design, the County Class missile destroyer, moved from making modern warships to reconstructive archaeology. Of his subject, the Athenian trireme, no physical remains have been found.[16] What can be gleaned from the literary records of performance, pictures on pots and the traces of ancient dockyards is painstakingly pieced together to reveal the technology of the time. This is effectively a design process, achieving a completeness and fit in technical and cultural terms that is seldom matched in the helter-skelter of contemporary procurement. Obviously it is impossible to undertake such exercises uncontaminated by modern preconceptions, but this archaeological method offers a design approach enriched by historicism.[17] Undercurrents referencing the past are inevitable in all high art and are sometimes overtly conjured up. The British architect John Young, a partner of Richard Rodgers, in striving to produce seminal modern housing by the river Thames reproduced the drawing style of Pierre Chareau (1883–1950), author of the astonishing Maison Verre, Paris (1932). (Chareau is particularly celebrated as an isolated and heroic proponent of the modern movement, his minimal oeuvre having great leverage.) Young's team filed patent tubular pens to match the line quality of early machine-made steel nibs. The meta-system of the draughting tools would somehow directly influence the outcome of the design work.[18]

The designers of the Royal Victoria Dock footbridge (1996), in London's dilapidated wharf area, reacted strongly against the prevailing fashion for iconic bridge

12.8a and b
**Royal Victoria Bridge, London Docklands, 1998**
This modern version of a transporter bridge was developed as an inverted fink truss assembled without bolts or welding so as to be fully demountable. The design was conceived as being made from found objects, the spars and wire rigging of the steel grain clippers that once used to dock there.

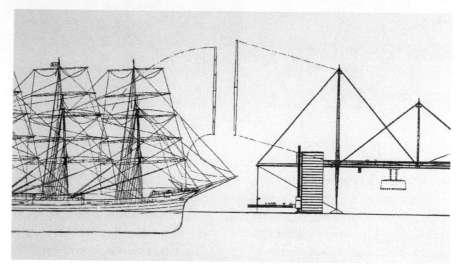

structures presented as catalysts for urban renewal. Their proposal to update a superannuated system, the transporter bridges of Ferdinand Arnodin (page 142), took a grammar of imaginary objets trouvés: the spars, steel strand and dockyard riggers supporting the grain clippers that once docked there (**figure 12.8**). Out of these bits was assembled a contingent construction, held together without bolts or welding, ready for dismantling, redolent of surroundings long gone.

## Regional engineering

The careful restraint exercised in the work of Jürg Conzett (1956–) and others from around Chur in the Swiss Graubund has produced structures of quiet, unprecedented beauty and seeming ease of execution (**figure 12.9**). Following the precepts of earlier Austro-Hungarian designers such as Adolf Loos (1870–1933),[19] whose insistence on the moral dimension of building leads directly to an attractive austerity, Conzett has

developed very specific ideas through a succession of small bridge and building projects. Hierarchies of simple systems are stripped of their superfluities and recombined into spare and refined wholes where every part contributes fully. An early footbridge across the Viamala Gorge is a complicated exercise in building up simple timber baulks and standard galvanised tie wires.[20] Robert Maillart's conceptual separation of supporting funicular line and stiffening beam becomes at construction stage a physical partition enabling the structure to be prefabricated and delivered in two parts by helicopter. A subsequent bridge in the same area reinstates the old Roman road as a long-distance footpath. The deck is treated as a simple set of stones suspended in space. Straps of high-strength duplex stainless steel carry the pavoirs in a shallow sag across boulder chokes and whirlpool-cut rocks. Panels and support strips are stressed against one another to make the bridge rigid. Elemental balustrades draw the parts together. Much of the pre-stress put into the bridge to lock its voussoirs together, expressed in the wedges and jacking hasps on the strap ends, has now been lost and Conzett relies on Othmar Ammann's small deflection theory of suspension bridges for the stability of his design.

## Modern lightweight structures

To date, the recipients of architectural awards for engineering achievement have tended to be structural designers rather than environmentalists. Frei Otto received a Gold Medal from the Royal Institute of Architecture in 2005, his contribution to building design considered revolutionary. The German Jörg Schlaich (1934–) has carried forward

12.9
**Pùnt da Suransuns, Graubunden, 1997–1999**
The ancient Roman road towards Dalmatia is carried through space on minimal straps of stainless steel: infinitely durable. A fully developed understanding of suspension behaviour, localised stress concentrations and pre-stressing are essential to achieve the apparent simplicity. Joints are seen as potential weak points, to be avoided if possible.

12.10
**Courtyard roof, Museum of Hamburg History, 1989**
Minimum-weight surface structures relying on their
double-curved forms for strength and stiffness fascinate
contemporary engineers. Computer analysis makes
formerly intractable problems readily soluble, so design
effort has moved to process and detail. The free form of
this roof effaces its impact on the rectilinear courtyard.
Junctions and glass fixings reduce to sheared and drilled
plates.

Otto's work and teachings to make tension structures, cable-stayed roofs and
suspension bridges of extreme sophistication, all the time referring back to a philosophy
of sustainable use, husbanding the world's resources. Lightweight structures are seen
as offering a legitimate use of finite material and energy supplies.[21] Re-configurable and
potentially devoid of waste, minimal structures of high-technology material have been
offered as a solution to problems as wide-ranging as disaster relief and historic building
renovation. The roof of the Hamburg Landesmuseum is a fine example of a glass
meniscus set over a medieval courtyard (**figure 12.10**). Conceptualised by tension
analogue, the braced compression grid is accessible to the same programmes
developed for tensile form-finding. The success of the practical construction relies on a
simplicity of detailing, reducing the parts, junctions and surrounding roof penetrations
to components of the crudest manufacturer. Another follower of Frei Otto, Werner
Sobek (1953–), has finessed this rigorous reduction of embodied complexity in a series
of glass façades and double-skin building envelopes. He re-connects with the use of
configurations and assembly details serving more than purpose.

The Deutsche Post building in Bonn (2002), by architect Helmut Jahn
(1940–), is a well-developed contribution to the polemics of energy use, a government
building made on behalf of Germany's green lobby. The limitless question of just how
we are to inhabit the planet and use its resources with future generations in mind is
met with the design of an uncompromising glass skyscraper. North European light is
let deep into the plan. Sobek develops the envelope as buffered layers of suspended
glass, simply clipped and squeezed into the finest wafers. Ventilation provisions allow
heat to be collected in or released away from the sunlit façades. These men have an
utter confidence that the material products of modern industry will maintain our
future. Ever higher technology remains intrinsic to human development.

## Purity in engineering

Further east in Europe, structural engineering has retained something of its tradition
for technical purity. It is almost as if the engineering of things keeps getting harder,

ever greater complexities of mathematics seeking diminishing returns. Specific construction systems have been focused upon and developed in great depth. Decades of enforced theorising in an economy where little got built has allowed one Czech engineer, Jiri Straski (1946–), to perfect a specific kind of bridge structure, the so-called stressed ribbon. A long time after the taxonomy of bridges seemed complete, this form with its particular beauty, combining elegance and economy, has emerged. Only when materials scientists produced tendons of sufficiently high strength could the sag of suspension structures be pulled out straight enough to walk on directly. The idea originated in the work of Ulrich Finsterwalder (1897–1988), of the Diwag bar company. He was tasked with looking for applications of the manufacturer's new high strength steel tendons and brought pre-stressing and pre-casting techniques together to make a bridge the deck and supporting structure of which is seamless. A long process of finding details to overcome concrete shrinkage and control lateral stability and dynamic behaviour followed, which have layered vestiges of other bridge types into the new form.

## Globalisation and regionalism

In the face of burgeoning globalisation, the dialectic is maintained by a resurgence of the importance of place. Air-conditioned tall buildings have become almost ubiquitous, the favoured symbol of economic progress.[22] Can they be a means of re-introducing a new vernacular? On the west coast of North America, the concerns of Europeans for energy conservation and structural expression seem to have been initially appreciated then almost at once formally dispensed with. The super-frame concepts of Illinois engineers (see page 176), influential in Europe and the Far East in such buildings as I.M. Pei's Bank of China building (1990) in Hong Kong or Foster's Commerzbank in Frankfurt (1997), have been passed over in favour of the steady enlargement of more easily constructed trabeated frames. In its pared-down state, with high-strength steel inserts and dry fire casings, the structures of the World Trade Centre in New York (1973) could not sustain the appalling terrorist attacks of 11 September 2001. The notion of hardening large building structures against such action shades into the Faustian.

## Corporate engineering and individual responsibility

The limits of personal responsibility in a technological age, well recognised in contemporary times,[23] do not seem at all diminished by the new scale of buildings. The saga of New York's Citycorp tower highlights some of the timeless and tragic aspects of engineering's nature. The very experienced designer William Le Messurier (1926–2007) completed a major scheme incorporating a novel podium-level transfer structure,[24] fully checked and approved. Important joints were changed from welded to bolted connections, easing construction but reducing robustness. A bright student became concerned about a failure mechanism, a 'twisting sitting-down'. A phone call precipitated the realisation that a hurricane could cause the complete collapse of the fifty-nine-storey tower. The client was informed; the public were kept in the dark.

Clandestine repairs were initiated, with welders working out of hours to conceal the repairs from the tenants. After three months and with plating-up only half complete, Hurricane Ella blew in towards the city. Evacuation plans for downtown were set in motion but the storm centre veered away. Reputations had been at stake and the temptation to prevaricate a very real test of character. The story was kept secret for twenty years.

## Wind engineering

The improvement in design tools is currently the most important agent of change within structural engineering. The make-up of these artefacts contains their own histories. The development of wind engineering has been heavily coloured by its transference from the aeronautical industry. Laminar flow models and testing facilities, relevant to aerofoils, were seriously deceptive and conservative when applied to bluff buildings on a jumbled ground plane. As buildings and clusters reached sufficient size to influence airflows around them and so require aerodynamic study, there was no revolutionary revision to aeroplane designers' formulations. Instead boundary layer theory was 'bolted on' to building engineering to make a rather Heath-Robinsonish canon of knowledge. Contemporary wind engineering is chiefly distinguished by the necessity of knowing the origins of its parts for safe application, a characteristic of benefit only to the specialist consultant.

## Designer's tools

Many of the new implements suit the generalist, perhaps too much so. The finite element revolution is nearly but not quite complete. Three-dimensional computer drawing packages can now model a proposal, all its parts and material properties. Developed initially to check for clashes in space, systems have now moved on sufficiently to record the contact between things. At that point it becomes easy to transcribe the data as a structural idealisation, put it in a load environment and check its adequacy. Beautifully post-processed images confirm whether the thing is working.

## Parametric modelling

Another step leads to parametric modelling. Relationships between parts are set up and the object file is bounded by certain fixes. As a design is manipulated on the screen, critical dimensions are maintained automatically. A stair riser will be added as floor heights are varied, windows stay appropriately spaced as an elevation is re-proportioned, sight lines are re-confirmed as an auditorium or stadium design is altered. By identifying stress, strain and buckling coefficients as key quantities, the practicality, even efficiency, in monetary or embodied energy terms of a structure will be instantly updated and maintained as schemes are drawn and manipulated. The engineer is in the box. Once learning systems become commonplace, the 'ghost in the machine' will go beyond the limitations of the individual programmer's knowledge.

## Pro-active structures

Is there any possibility of genuine structural innovation left to us? Real advances tend to be generated as escape solutions from technical impasses. The John O'Connell bridge across the Sitka Sound in Alaska (1972) is the first cable-stayed highway crossing in the United States (**figure 12.11**). Its environment is harsh and the structure quickly showed excessive response to the wind. The movement was not just the aerodynamic interactions previously described (see page 180), vibration-driven by air flows forced into pulsating patterns. Instead, buffeting was building dangerous energy levels in the brittle cold.

The wind at any place on the earth's surface can be described as a 'power spectrum', a signature composed of different pressure frequencies. There is the annual change of the seasons, the weekly variation of the passing weather systems and the staccato gusts and chops induced by surrounding topography. Each peak carries an intensity of energy. In turn, every structure comprises a hierarchy of forms, contiguous or loose-bonded. The response spectrum of such assemblages is another signature, the frequencies at which the whole and its parts will oscillate naturally.

When two such systems are put together and then matches are found across the profiles, the structure is vulnerable. Buffeting will produce a large dynamic component of load over and above the steady pressure of the wind. The addition of stiffening and increasing or redistributing weight is expensive, and at Sitka retro-fitting additional strengthening elements was barely feasible. Active control was considered as a remedy and a beautiful study in 'pro-active structure' resulted.[25] If actuators are added to a dynamic system then its characteristics are modified. At relatively low levels of intervention, external excitations can be countered before they grow to unmanageable levels. Hydraulic jacks, rams or piezo-electric materials straining under electrical impulse are controlled by a processor responding to sensors out on the frame. If the feedback control is correctly tuned then the merest trickle of power will keep the lightest of structures aligned in the harshest of conditions. So far such applications have only been adopted in extremis. Active damping systems are installed in a number of towers along the Pacific Rim waiting to act against earthquake. Particular classes of industrial installation, pipe bridges and aerial pylons, have proved ideal sites to study the practicalities of such systems.

Perhaps one day, when total energy use, embodied and service requirement, is truly recognised as the cornerstone of our future there will come a new class of buildings, shimmering towers and enclosures imperceptibly shifting and straining in response to patterns of circulation, sun, rain and wind.

This ends pessimistically. The structural engineer's tool box is full. A palette of new materials beckons. Problems suffer fewer constraints so solutions have less definition. What is being called for here is a stronger awareness of the influences that affect the way we construct things, a recognition of the feathered edges of our field of enquiry. This is an intermediate step to a new engineering. The objective is to authenticate technology's contribution to our culture.

12.11 (*opposite*)
**John W. O'Connell bridge, Sitka, Alaska, 1971–1972**
The first cable-stayed bridge in North America proved prone to aerodynamic instability.
A detailed proposal was worked up to control the movement of the bridge with an active control system; hydraulic actuators guided by sensors would respond to deflections to keep the structure steady. Although not implemented, the idea and system detail presage a new era of ultra-efficient structures reacting mechanically to the surrounding load environment.

# Notes

## Introduction

1   The Burgess Shale is a stratum of Precambrian rock containing a profuse fossil record of weird animal life.

2   Spindler, 2001. 'The Man in the Ice' is a chalcolithic (copper age) mummy of inestimable importance to our understanding of prehistoric technology. He was carrying an unfinished yew bow, revealing how it was made. His mixed bag of copper and stone implements show how technology is always transitional, old co-existing with new.

3   Subrahmanyan Chandrasekhar. 'Newton and Michelangelo'. *Meeting of Nobel Laureates*. Lindau. 1994.

4.  J. Valerie Fifer, *The Master Builders: Structures of Empire in the New World-Spanish Initiative and United States Invention* (Durham).

## Chapter 1

1   The 'three age system' was an idea alluded to in antiquarian writings since the beginning of the nineteenth century. The classification was given shape by Christian Jurgensen Thomsen (1786–1865), curator of the National Museum of Denmark. Forcing technological development into a chronology obscures much more complicated processes of advance and regression.

2   Thom, 1971. The theories of Professor Alexander Thom (1894–1985) have been fiercely contested by many archaeologists. The precision and purpose of the rings he studied may have been misinterpreted but the idea that the layouts had astronomical significance and influence upon social organisations is generally accepted. That such an intense concentration of resources would have been for specific practical purpose is sound engineering logic.

3   Chalavoux, 2001, p. 9. The 'golden section', the ratio of the whole to its part remaining constant, appears in many growth patterns. Natural objects appear well balanced but the exact ratio cannot be detected by eye alone. In ancient times its use may have been an aesthetic effort to create a whole related in all its parts or it may have appeared spontaneously when dimensioning things using proportions only.

4   Tyldesley, 2003. The first real building structural engineering problems – proving foundations, taking materials towards their practical limit (vault voussoirs), dealing with instabilities (slope failures), and ensuring longevity without maintenance – were all met in a civil engineering context, artificial hills, rather than in habitable construction.

5   Verner, 2002. It turns out that the resolution of the technical problems associated with the construction of big stone mounds followed a rapid, experimental trajectory driven by a momentum of ideas and by individual will. Why was it ever thought to have been a long evolutionary accretion of custom and precedent?

6   Parkinson, 1999. The notion that building technology is a carrier of cultural information and is not therefore referred to in other cultural productions would explain the historical 'invisibility' of engineering as an index of a society's condition.

7   Verner, 2002, p. 165. Accretion layers were a common form of construction suited to making a tomb as large as possible yet ready for an uncertain death. This is not the soundest structure. The Meidum pyramid (2600 BC) had catastrophic design flaws. Although the weakness of in-built slip-planes could

be offset by cross bonding, some of its internal joint planes were polished smooth. Blocks in the core were tilted inwards, improving strength, but the outer layers (now collapsed) were laid flat.

8   Verner, 2002, p. 74. The south-east corner of the Great Pyramid of Giza (2467 BC) is two centimetres higher than the north-west corner. The prevailing wind from the north-west could raise the surface of a water trough that amount over such a length.

9   Verner, 2002, p. 312. The tiered arches over the burial chamber in Niuserre's pyramid incorporate several structural nuances to resist earthquakes and superimposed pressures. A compromise shows between the labour involved and robustness achieved.

10   Giedion, *Architecture and the Phenomenon of Transition: The Three Space Conceptions in Architecture*, 1952, p. 2. Starting from the compression of space in Egyptian hypostyle halls, in excess of the limitations imposed by technology, Giedion argues for a conception of space completely alien from but neatly opposed to the modern sensibility.

11   Pliny, *Natural History*, Book 36.14. A canal was dug from the river Nile to the spot where the obelisk lay and two broad vessels, loaded with blocks of similar stone a foot square – the cargo of each amounting to double the size and consequently double the weight of the obelisks – was put beneath it, the extremities of the obelisk remaining supported by the opposite sides of the canal. The blocks of stone were removed and the vessels, being thus gradually lightened, received their burden.

12   Jones, 1995, p. 65. Carved reliefs demonstrate that the big obelisk barges were considered impressive. Carrying two stones on one hull demonstrates confidence. The detailing shows vast size achieved both by scaling up and multiplying proven systems.

13   Willetts, 2004, p. 57. Ancient Cretan settlements were unfortified and traces of cultural exchange are all around. The high cost of bronze implements ensured stone's persistence, and stone axes appeared patterned to emulate metal.

14   Popper, 1998, p. 33. Xenophanes is credited as the founder of the Greek Enlightenment: nature approached through geology and meteorology, humanity through literary criticism and social theorising, searching for a theory of knowledge. His ideas were considered lightweight in comparison with other thinkers' cosmologies and the final steps to the western Enlightenment had to wait.

15   Plato, *Meno*, 80d. The idea that knowledge must be innate to the soul implies certain characteristics. Things are veiled. They are not synthesised but uncovered. Genuine knowledge must be timeless. All these properties remove engineering from the immediacy of action.

16   Plato, *Timaeus*, p. 21. 'And every sort of body possesses solidity, and every solid must necessarily be contained in planes; and every plane rectilinear figure is composed of triangles; and all triangles are originally of two kinds, both of which are made up of one right and two acute angles; one of them has at either end of the base the half of a divided right angle, having equal sides, while in the other the right angle is divided into unequal parts, having unequal sides.'

17   Coulton, 1977, p. 15. Architects two millennia ago were asked to combine a theoretical knowledge with the practical organisation of the building site. Despite rigorous conventions, design was still expected to provide nuance and idiosyncrasies.

18   Kenny, *Ancient Philosophy*, 2004, p. 41. The historical Socrates needs reconstructing from the records of others. His reasoning from the general to the particular sets up one kind of engineering – the contingent. Finding applications from theoretical discoveries is another, quite different, approach.

19   Cuomo, 2001, p. 43. Mathematics has then a double character. There are two kinds of arithmetic: that of the 'many', the money-oriented traders and merchants, and that of the 'thinkers', people who philosophise.

20   Singer et al., *A History of Technology*, 1954, vol. 2, pp. 629–657. Archimedes was a geometer because geometry is a technician's science. His researches on statics revealed the fundamental principles relating to the lever and the centre of gravity. His studies were of great assistance to those who construct purposeful machines.

21   Monk, *Bertrand Russell*, 1996, p. 25. Bertrand Russell's career was initiated by his encounter with and fascination for Euclid's *The Elements*. He wrote, 'I had not imagined that there was anything so delicious in the world. It was as dazzling as first love.'

22 Monk, *Wittgenstein*, 1991, p. 26. As a teenager Wittgenstein read *The Principles of Mechanics* by Heinrich Hertz, in which the term 'force' is queried. Hertz's proposal to remove the 'painful contradictions' at the heart of engineering by recasting Newtonian mechanics without the concept of force at all appealed to the young thinker, furnishing him with a model for solving philosophical problems.

23 Coulton, 1977, p. 55. Greek architects worked with *paradeigma*, specimens, to achieve uniformity. The word denotes a mythological example used to underscore a literary point, in Homer often modified to appear real-worldly. Connotations of nature amplified and ideals realised attach themselves to the artefact.

24 Humphrey, *Pausanias: Description of Greece*, 1997, 5.16.1. 'It remains after this for me to describe the temple of Hera and the distinguished objects in it. The temple's style is Doric, and columns stand all around it. In the back room (opisthodomos), one of the two columns is made of oak (the rest are stone).'

25 Humphrey, *Pausanias: Description of Greece*, 1997. 'The treasury of Minyas, a wonder second to none in Greece itself or elsewhere, was constructed in the following manner: it was built of stone and its appearance is rounded since its crown does not rise to too sharp a point. They say the highest stone is the keystone to the whole structure (an early example of a misconception, probably laid to conceal craft knowledge). The fortification wall of Tiryns, which is the only vestige of the ruins left, is a work of the Cyclopes and is made from unworked stones, each stone being so large that the smallest of them could not be moved even in the slightest degree by a pair of mules. Long ago, smaller stones were fitted in so that each of them is tightly fitted with the larger stones. (Mueller shows how Cyclopean masonry was economic and exceptionally robust, locking together and redistributing load paths without collapsing in earthquakes).'

26 Braudel, 2001, p. 117. The Greek term *anachoresis* describes the wider process of moving on. Colonisation and emigration, trading, raiding and long-distance incursions were all aspects of a particular Greek attitude to mobility.

27 Starr, *The Influence of Sea Power on Ancient History*, 1989, p. 4. Historians, notably Alfred Mahan, allotted the Greek Trieres a pivotal role in shaping the early history of the Mediterranean basin, an argument appearing just at a time when a modern maritime arms race needed justification. Starr re-evaluates this technical determinism in the wider context of economic and social conflicts.

28 Morrison, 2000, p. 2. Acknowledged by historians as the pre-eminent weapons system of the ancient world for well over a century, it has been argued that only by reconstruction can the characteristics of an Athenian trireme that gave such vessels staying power yet caused their eventual supersession be identified.

29 Morrison, 1986, p. 208. A persistent failing of warship design has been the sacrifice of robustness for speed: the battle cruiser losses at Jutland; the Italian 'cardboard cruiser' *Bartolomeo Colleoni* rapidly dispatched at Cape Spada. The reconstructive archaeologist John Coates contributed to a very lightweight destroyer design for the British Labour government. The concept of a thin aluminium superstructure was found wanting in the Falklands War. The recreated trireme, influenced perhaps by rowing eight design, may be a little lightweight compared with the originals.

## Chapter 2

1 Goldsworthy, *The Punic Wars*, 2001. The Republic was aware of the consequences of subsuming Carthage and many senators argued against it. As the wars were won, technology was transferred, economic balances changed and Rome was irretrievably set on a road to empire and subsequent dissolution.

2 Galinsky, *Augustan Culture*, 1998, p. 333. Augustus' use of Hellenism was characteristically subtle; an appropriation of classical motifs to underscore 'Republican' credentials and a reinterpretation of forms, such as the introduction of the Corinthian order, to brand the new age.

3 Plutarch, *Life of Numa*, xix–xx, seventh century BC. It was accounted not simply unlawful, but a positive sacrilege, to pull down the wooden bridge. An oracle had required it to be made entirely of timber without iron fixings. It was eventually replaced in stone in the time of Aemilius Quaestor (AD 126).

4   Caesar, *The Gallic Wars*, Book IV, 17: 'The greater the force and thrust of the water, the tighter were the balks held in lock.' He seems to have had a structural action explained to him and to have only partially understood. The timber pairings above and below each crutch, which would be necessary for a scissor effect to tighten the assembly under lateral pressure, are not mentioned, contrary to so many other clear descriptions.

5   An 'ovatio', a lesser and non-military triumph, could be granted for a particularly outstanding example of civil engineering or public building work. The first pantheon by Marcus Vipsanius Agrippa constituted an 'ovatio' for the princeps Octavian.

6   Humphrey, 1998, pp. 36, 104–106. As well as reflecting the Republican virtues of utility, it is the robustness of the channel's construction, unyielding to flooding or backwash from the Tiber, that the author celebrates. The various construction campaigns that made up the Roman sewerage network can be traced in the variety of materials used.

7   Galinsky, *Augustan Culture*, 1998, p. 200. The characteristic synthesis of reforming experimentation and appeal to tradition was well met in the Augustan adoption of the Corinthian order. Those in the Forum Augustum were modelled on late classical and fourth-century models but with tectonic and ornamental components re-balanced in a new way.

8   Pearson, *The Roman Shore Forts*, 2002. Just as the Americans left their aircrafts shiny and unpainted towards the end of the Second World War, so the Romans displayed their dominance of the Mediterranean with bright burnished vessels. In the English Channel they were not quite so confident.

9   Vitruvius, *The Ten Books of Architecture*. Hailed as one of the most important books on architecture, this handbook was written by an artillery man in the Roman army. The text gives equal weighting to technical considerations and aesthetics.

10  Taplin (ed.), *Literature in the Roman World*, 2000, p. 39. Handbooks for practical purposes appeared in the last century BC in response to real concerns over control of empire and potential decline: 'Cicero attempts to forge a stable language to bolster an increasingly unstable reality.'

11  Humphrey, *A Sourcebook: Dio Cassius, Roman History*, 1997, 68.13 1–6. 'Trajan constructed a stone bridge over the Danube ... At present only the piers are standing ... since Hadrian removed the superstructure, fearing that if the guards at the bridge were overcome by the barbarians there would be an easy crossing into Moesia.'

12  Ball, *Rome in the East*, 2001. Gallic cavalry captured at the battle of Carrhae ended their days guarding the eastern frontier of the Parthian empire, modern day Turkmenistan. When the Han Chinese overran this area, these Romans became their subjects in turn.

13  *Art Encyclopaedia*, Sassanid Design, 1968 McGraw Hill, pp. 12–704, p.397. Sassanid structures display the interplay of forces within them. Their static balance arrests space, in complete contrast to Roman experiments with dynamic spaces and concealed framing.

14  Robert Byron, 2001. Byron describes the bricks as ill-fired and marvels at their survival over fourteen centuries of weathering

15  Klaus Dunkelberg, *Institute of Lightweight Structures, IL31: Bamboo*, 2005. The ideal curve of a self-weight arch is a catenary, a transcendental curve. A wand twisted at one end and tied to a neighbour takes up a parabolic profile. The armature of a mud hut must be sufficiently flexible for the shell to move towards its best shape as it is moulded up by hand.

16  Watson, *Aurelian*, 1999, p. 2. Labelling the transition from classical to late antiquity as a 'crisis' oversimplifies an obscure period. A strong switch towards cheap impermanent construction hampers the archaeological record.

17  Cuomo, 2001, p. 86. Philo of Byzantium, in his treatise *Belopoiika* (war engines), described a trial and error process to which a mathematical formulation was fitted. It's not bad. The pivot pin diameter is cubed to set up the sizes of other parts of a ballista. He compares this process to the method used by sculptors proportioning their work.

18  Lindsey Davis, *The Jupiter Myth*, 2002. Recent archaeology reveals that the Romans never took London bridge to the stage of 'permanent infrastructure'. This says much about their attitude towards the occupation of Britain.

19   Alex Butterworth and Laurence Ray, *Pompeii*, 2006, p. 36. Ancient Rome came to rely on southern Mediterranean imports, and pozzolanic cement served as ballast for the outward journey of Roman grain ships to Egypt. It could be used at destination to improve harbour facilities. The modern use of concrete retains something of that early exploitation of the low skill-levels of remote and subject peoples.

20   Yvette Goepfert, *Pont du Gard* (Editions AIO), 1989. Aqueducts obviously needed careful grading. Arches were either close-spaced to reduce flexibility or set in diminishing tiers approaching the final level. The Flavian amphitheatre is typical of the massive structures that were erected on poor ground. Not so intolerant to movement, the tiered construction of these buildings helped achieve even cornice lines.

21   Vaquerizo Gil, *Guia Arqueologica de Cordoba*, 2003, pp. 23–26, 50–51. The Augustan temple of Diane in the provincial forum of Cordoba, Spain, has an exposed plinth and foundation system visibly bigger than the delicate Corinthian colonnade above it.

22   Werner Muller and Gunther Vogel, *Atlas zur Baukunst*, 1987, p. 301. Polygonal masonry minimises labour. Stones are simply bruised down to square faces then assembled into an interlocking pile. The variety of faces all transfer load and the patterns change as the blocks shift.

23   John Julius Norwich, *A History of Venice*, 1989, p. 6. Besides defensibility, the commodity of salt and the imperative for seagoing expertise made an improbable beginning the most romantically charged commercial success story in history.

24   G.P. Baker, *Constantine the Great*, 2001, p. 161. The Emperor's adoption of the Christian religion is interpreted as a new way of understanding and controlling society rather than imposing upon it. This is a momentous recognition of the representative quality of intellectual artefacts.

25   Xavier Barral I Altet, *The Early Middle Ages*, 2002, p. 39. Pilgrims would visit these artificial catacombs, circulating around internal colonnades and looking down into the man-made grottos of the crypts.

26   Cameron, *Procopius and the Sixth Century*, 1996, p. 86. The emphasis on the panegyric element of the text obscuring information on ancient building methods provides literary analysts with their raison d'être. Modern reports tend not to dwell on difficulty.

27   'Cistercian Architectural Purism', *Comparative Studies in Society and History*, vol. 3, no. 1 (Oct. 1960), pp. 89–105. The reasons for austere prescriptions are easy to see. It is also fascinating to trace to what extent such perfection could be achieved and maintained.

28   Xavier Barral I Altet, *The Early Middle Ages*, 2002, pp. 99–117. That high industrial technology develops concurrently but only in parallel with imperialism means that when re-appropriated it is drained of any possibility for cultural coding. Visigothic under-exploitation is the classic example of this phenomenon, which also accounts for the barrenness of modern globalism.

## Chapter 3

1   Torvald Faegre, *Tents: Architecture of the Nomads* (1979). The variegation of tent types can be put down to different environments, customs, power structures and subsistences. 'Structural' features held in common can be uncoupled from these influences.

2   Conrad Totman, *A History of Japan* (2005), p. 47. The distinctive character of the country is made up of many parts, a unique geography and profound history.

3   Kiyosi Seike, *The Art of Japanese Joinery*, 1977, pp. 7–10 (Translator's introduction). A process of repeated revolution then consolidation is described, perhaps an inevitable consequence of the problem of building but also a pattern found in other discourses in Japan.

4   Hideo Sato and Yasua Nakahara, 2000. A grasp of Japanese structural timberwork can only be attained by close study of the individual joint types and attitudes of the carpenters themselves.

5   Wai-Fah Chen and Charles Scawthorn, *Earthquake Engineering Handbook* (2003), 19.2–3. Pagodas rely on a variety of mechanisms for their earthquake resistance. There are many levels and types of shock.

6   Henri Deneux, *L'Evolution des Charpentes du XIe au XVe Siècle* (1927). This surveyor of the French Historical Monuments Service recognised the patterns of jointing across more than five hundred examples.

7 There are natural means of manipulating timber sizes. Underplanting oak with ash forces the growth straight and high in competition.

8 J.A.S. Evans, *The Age of Justinian*, 2000. Procopius was writing a flattering record but his dates (23 February 532 start, 26 December 537 complete), seem precise enough. The project must have been in planning before the Nike revolt but that incident put an onus on speed of construction, faster than the mortar could set and perhaps design decisions could be safely taken. The first dome partially collapsed on 7 May 558.

9 Rowland Mainstone, *Hagia Sophia: Architecture, Structure and Liturgy of Justinian's Great Church* (1988), pp. 85–128. Cracks, repaired cracks, remedies and misalignments are all plotted out to show how the structure has lived and breathed.

10 Sir Banister Fletcher, *A History of Architecture on the Comparative Method* (1948). Another of Justinian's churches, this small building is beautifully sited on the shores of the Bosporus with dome and pendentives, just like at Santa Sophia, but the dome with peculiar ridging increasing its stiffness.

11 Nikolaus Pevsner, *An Outline of European Architecture*, 1991. This is a classic (effusive and reductive) example of a western spatial analysis of a monument which shows many complicated influences.

12 Banister Fletcher, 1948, p. 251. The beautiful drawings record the use of standard amphora as void formers to lighten the concrete shell but also special interlocking cylinders in the upper section of the dome.

13 Salt towers. The transport routes across middle Europe of the lucrative medieval salt trade required protection. A special kind of isolated fortified watchtower appeared: tall light superstructure over masonry base.

14 Dimitri Gutas. *Greek Thought, Arabic Culture*, 1998. The nature of the interchange between East and West remains an intellectual football. The Arab world did not just store Greek philosophy for the West to re-address when ready but of course used it for its own purposes, and many western treatments overlook this. The presentation in these pages fails in that respect and it needs addressing elsewhere.

15 Sir Thomas Heath, *The Works of Archimedes* (2002). He may have invented the screw, but the lever is older. Archimedes made a clear exposition of how and how much mechanical advantage a lever can give.

16 Boyer, *History of Mathematics*, 1989. The use of symbols had Indian precedents but was finally separated from number theory by the Persian astronomer Al-Khwarizmi (780–850).

17 Levi, *Renaissance and Reformation*, 2002, p. 36. The re-introduction served to break the Scholastics' reliance on revelation and, through the dilutions of translation, opened a resistance to canonical interpretation.

18 Fletcher, *Moorish Spain*, 1992, p. 150. The Islamic 'translation movement' and the appropriation of its work by western universities left plenty of room for original texts to be made more palatable. The use of the texts short-circuited accusations of heresy and allowed Orthodoxy to move along.

19 Nuha N.N. Khoury, *The Meaning of the Great Mosque of Cordoba in the Tenth Century* (1996). The physical remains of Roman and Visigothic buildings are incorporated in the new fabric. As architecture in exile, the design was also more permeable to the motifs of previous occupiers.

20 Gernet, *A History of Chinese Civilisation*, 1996. The policy of appeasement was a subtle economic tool, productivity channelled directly to defence. Long-term stability made room enough for invention.

21 Circassian iron, the name ancient Romans gave to the first steel they encountered. Heat treating metals was a speciality of the region back to the seventh century BC.

22 Joseph Needham, *Science and Civilisation in China*, 1954, pp. 201–202. Several poems about chain bridges allow their geographical spread to be charted. The Lu-ting bridge, site of a decisive skirmish during the Long March, graces a postage stamp.

23 The famous Zhaozhou bridge built between 595 and 605 has a low arch ring of limestone slabs cramped with iron and open spandrels to reduce top weight. Surviving several strong earthquakes, it has served as a pattern scaled up or down to suit many other sites.

24 *Secrets of Lost Empires: China Bridge*, ITV, November 2000. Rainbow bridge. Using an image from a nine-hundred-year-old painting, a wooden canal bridge was re-created for a television documentary.

Details were speculated upon and tested. The small scale of the project was not unsuited to an agricultural village's resources or the programme's running time.

25 Dajun Ding, 'Ancient and Modern Chinese Bridges', *Structural Engineering International*, vol. 4, no. 1 (February 1994), pp. 41–43. Corbel bridge. Simple timber cantilevers are superimposed on one another. The assembly is an intermediary stage in the development of the dougong (corbel bracket), a feature of Chinese architecture which first appeared in the Western Zhou dynasty (1066–771 BC).

26 Valentin A. Sokoloff, *Ships of China*, 1982. The myriad variety of vessels all reflect their functional requirements and environment: fishing vessels of oiled timber subject to icing in northern waters, cargo boats with a specific curvature to suit the waveforms of the Formosa channel, even a junk to ride the river bores.

27 Hasler and MacLeod, *Practical Junk Rig*, 1989. Originally concoctions of bamboo matting on stiffening battens, the varieties of balanced lug sails and fans of trading and fishing vessels closely fix their origin.

28 Thorlief Sjovold, 1979, p. 34. Strikingly high prow and stern combined with low freeboard, very fine construction and ornamentation and a comprehensive range of 'accessories' suggest the modern designation of yacht, a state or pleasure craft.

29 Stave churches. The churches were framed with beams, posts and vertical plank infills. The old Norse for bearing post, *stafr*, gives the name.

30 Angelo Forte et al., *Viking Empires*, 2005. With a beautifully developed maritime technology, the Vikings felt little need to invest in other improvements, relying on raiding and borrowing other technologies, tactics and customs. Their trestle bridges across marsh and glacial lake match patterns found eastwards.

31 Eric Christiansen, *The Norsemen in the Viking Age*, WileyBlackwell, 2002, pp. 84–85. '"Trelleborg problems" have not all been solved, least of all the big one: what were they the forts for?'

32 Drachma. The Athenian 'handful of nails' comprised six obels, iron rods that could be grasped in the hand.

33 Rosamond McKitterick et al., 1999, p. 185. The straightforward development of flat-bottomed Celtic barges, cogs displaced up to 200 tonnes, were robust and required a small crew.

34 Fred Pearce, *Deep Jungle*, 2005, pp. 160–161. The terraces, canals and colonnades of the Angkor Wat temple complexes are formalisations of agricultural patterns extending hundreds of kilometres around. Landscape engineering and architecture fix the jungle.

35 Heather Martienssen, *The Shape of Structure*, 1976, p. 53, The notion of a building remaining a 'cave' is given shape by excavating the living rock, leaving columns and pilasters to modulate the special feeling of space won from the solid.

## Chapter 4

1 Peter Brown, *Augustine*, 2000. The saint's equation of light with revelation came partly from the south Mediterranean light flooding through his study window and partly as an appropriation of rival religions' fundamental tenets.

2 C.H. Lawrence, *Medieval Monasticism*, 1984, p. 184. Bernard of Clairvaux, suffering permanent gastric disorder, ruined his health through the stricture of his own rule. Despite a penchant for mortification, his monastery plan was a mixture of convenience and practicality with austerity and severity.

3 Henderson, *Chartres*, 1968, pp. 81–102. A process of replacement is described which highlights designer's criticisms of the preceding work.

4 Sir Banister Fletcher, *A History of Architecture on the Comparative Method,* 1948, pp. 328–331, 383. The beautiful line drawings trace steadily increasing complexity and refinement in early cross vaults then jump cuts to stellar, fan and pendant vaults.

5 Richard Harris, *Discovering Timber-Framed Buildings*, 1993, p. 16. The practice of 'rearing' frames is held to have been relatively rare, as labour intensive. The short-term concentration of manpower did make larger elements more manageable.

6   Cecil Hewitt, *English Historic Carpentry*, 1997, pp. 161–163. The directionality of timber joints allows assembly sequences to be dissected. The octagonal lantern at Ely cathedral was designed so that the massive corner posts could be lifted without inclination so as to remain more manageable at height.

7   Gimpel, *The Cathedral Builders*, 1983, pp. 1–5. The fervour for church-building fits a Marxist analysis well. The workforce needed new social structures, and economic conditions allowed these to find form in an explosive expansion of building projects. Proletarian art appeared.

8   Courtney, *The Engineering of Medieval Cathedral*, 1997, pp. 169–190. Despite much hyperbole, the dimensions of Beauvais choir are not a magnitude above those of other cathedrals of the time and are proportionally very close.

9   Heyman, *The Stone Skeleton*, 1997. Demonstrably low stress levels in masonry are important if the stability of medieval cathedrals is to be treated as a geometrical problem. Inappropriate detailing and differential settlements can, however, generate significant forces across the stones.

10  Elliot Sober, *Simplicity*, 1975. An example of the more general principle of parsimony in nature and science, Ockham's razor is a tool rather than tenet and any justification will have an aesthetic component.

11  The Instituto of Structural Engineers, mission statement: 'Structural engineering is the science and art of designing and making, with economy and elegance, buildings, bridges . . .'

12  Norwich, 1998. Bernard promoted the Crusade on behalf of Pope Eugene III and became scapegoat for its failure. The excursion antagonised and united the Muslim world against the West.

13  Courtney, *The Engineering of Medieval Cathedrals*, 1997, p. 28. The point is well made that the words 'Euclid' and 'geometry' meant something very different to medieval masons than Euclidean geometry means to us.

14  Richard Southern, *The Making of the Middle Ages*, 1961, p. 193. Boethius linked geometry to arithmetic, music and astronomy. In translating Euclid's *Elements*, he has no idea of some of the terms.

15  Courtney, *The Engineering of Medieval Cathedrals*, 1997, pp. 27–53. The evidence shows that the medieval masons constructed their own 'constructive geometry' distinct from mathematical treatments with which to manipulate form. It was very simple.

16  Panofsky, *Gothic Architecture and Scholasticism*, 1951. Some things are beautifully proportioned, others are not. The notion of an ideal behind appearances began to erode beneath the variety and profusion of real life.

17  Padovan, *Proportion*, 1999, pp. 181–184. The controversy over the setting-out of Milan cathedral is well documented. Building within a triangle felt stable but generated irrational dimensions. The use of whole-number divisions instead implied completeness.

18  Ad quadratum. By rotating a square within a square, the irrational number root two appears and an area is halved.

19  Golding, *The Spire*, 1974. A description of the ground seething like a living thing under the foundations of a medieval church. The cut stones and geology both are both animate.

20  Norberg-Schultz, 1985. This curious affectation might be neo-platonic. Structure is being forced back under the surface of things, leaving only vestiges.

21  Arnold Pacey, *The Maze of Ingenuity*. In the course of wandering through the German mining regions, Paracelsus (Bombastus von Hohenheim) formulated a theory of social worth based on an extreme Protestantism which required a specific position to technology and the value of direct experience in moulding nature and making things.

22  Morrison, 'History of the Ship', *The Age of the Galley*, 2002, pp. 142–162. Stem and stern lines were modified, rounder and deeper, to give more flotation beneath the gun-carrying castles at either end of the heavier galleasses.

23  Greenhill, *The Evolution of the Sailing Ship*, 1996, pp. 225–232. These rigs lay far back along the evolutionary chain and the pictures from the time give much leeway to interpretation.

24  Braudel, *The Mediterranean*, 1996. The historiography suits the subject. The inland sea separates and brings together, and produces systems working over a wide range of timescales.

25  Brian Lavery, *Ship*, 2004. Usually displacing less than 200 tonnes, the lateen rig enabled the Xebec to run faster and point higher than any other vessel of the time, ideal for piracy.

26  Anthony Bonner (ed.), *Doctor Illuminatus: A Ramon Llull Reader*, 1993. The man's ebullience and curiosity for life comes across clearly in his writings. A troubadour at one time, his environment gave him plenty of scope to absorb and synthesise different patterns of thought into something new. He knew Catalan, Occitan, Latin and Arabic.

27  Ross King, *Brunelleschi's Dome*, 2001, p. 42. In lobbying for the job, the spectacular nature of his proposals helped Brunelleschi immeasurably in popular opinion.

28  Jeffrey Wigelsworth, *Science and Technology in Medieval European Life*, 2006. The crane is emphasised as the centre of the building site and was a construction of some substance and complexity.

29  Gould, 1999, 1.1. Leonardo's pagan belief in an animistic universe is well illustrated in his ideas of geology.

30  John Steer, *A Concise History of Venetian Painting*, 1980. Light conditions on the lagoon encouraged a particular treatment of paint. Valuable materials for colours, smalt and verdigris, passed through the port and into colourists experiments.

## Chapter 5

1   Renaissance. A 'discovery' of the Swiss art historian Jacob Burckhardt (1818–1897), the Renaissance was first detected in literature then revealed in culture and social institutions and finally traced to its economic and geographical origins. It is as much an artefact and demonstration of historiography as it is a definable event.

2   Lane, *Venetian Ships and Shipbuilders of the Renaissance*, 1992, p. 64. The 'technical renaissance' of the early sixteenth century gave Venetians enough confidence to reform their naval architecture.

3   Mari Biagioli, *Galileo Courtier*, 1993, pp. 19–30. Mathematics didn't appeal to court culture and stargazing did. Galileo's career (and programme of research) was guided by the success of his astronomical discoveries.

4   Lane, 1939, pp. 173–175. Each specialisation – gun-founding, sailmaking, chandlery – was arranged around a separate basin of the Arsenale. Raw material and partially completed products were towed from place to place, saving space and handling costs.

5   The earliest evidence is from a pottery ship's model found in a tomb from the first century AD. Chinese rudders were balanced on their pivot, a nuance that was ignored in the early western examples that appeared from about 1200 onward.

6   Timoshenko, 1983, pp. 11–15. The author is kind to Galileo and highlights his sound conclusions on scale effect. The errors in his beam theory show up in the discussion of Edme Mariotte's (1620–1684), correct formulation made some fifty years later.

7   Jerome Ravetz, *Scientific Knowledge and its Social Problems*, 1995. The social influences on Galileo's 'pure' science are well analysed in Ravetz's work. The mistakes made in Galileo's cantilever study are identified not as simple errors but as the outcome of forces at work around him.

8   James Lattis, 1994, pp. 195–202. The Jesuit astronomer Christopher Clavius tried all sorts of ways to protect then modify the Ptolemaic cosmology in the face of Galileo's discoveries. The arguments were not about alternative realities but over best explanation of the phenomena.

9   Addis, 1994. The rigid presentation of structural engineering development being through a series of paradigm shifts is tempered in Addis's subsequent writings by his close studies of contexts and continuities

10  Levi, *Renaissance and Reformation*, 2002, p. 208 The Renaissance is characterised by its multivalence. Self-determination of spirit shades into determination of nature. The human mind could have direct intuitive access to divine truth and was therefore perfectible.

11  Jozef T. Devreese and Guido Vanden Berghe, 2007. The famous diagram supposedly inscribed on his tombstone is an ingenious thought experiment to demonstrate equilibrium on an inclined plane. A rosary is draped over a wedge.

12  Boyer, *The History of the Calculus*, 1954, pp. 96–186. The concepts necessary for modern analysis gradually assembled themselves ready just in time for modern technology's demands.

13  Desargues. Another name recorded on the lunar landscape. The importance of astronomy to the development of mathematics is reflected in many of the titles given to the moon's lunar features.

14  Alberto Perez-Gomez and Louise Pelletier, *Architectural Representation and the Perspective Hinge*, 1997. With the recognition that representational systems have a direct influence on the conceptual development of designs comes the possibility of manipulating their effects.

15  Mozart, *The Marriage of Figaro*. Act 1 opens with the bass-baritone singing out the measurements of his marriage bed. All is harmony and perfect proportion about to be disrupted.

16  Venice Maritime Museum. A collection of original models in plaster show how surveyors worked up the planes and glacis of their fortifications.

17  Marin Getaldic. On the interface between West and East, well travelled and correspondent with many famous scientists of his day, Getalidic, Gheltaldus, Ghetaldi kept himself removed enough for original thought.

18  Yates, *The Rosicrucian Enlightenment*, 1972, p. 132. A printed tract entitled *Rosa Florescens* reflects well the concerns of the 'Rosicrucians'. Technical subjects were (apparently) in need of reform and re-connection with nature. The world makes up a sign system capable of revealing a deeper truth.

19  Ackroyd, *The House of Doctor Dee*, 1994. An attempt at recreating the consciousness of the early modern mystic and technologist.

20  Rybczynski, *One Good Turn*, 2001. The immense cost of hand-fashioning screws restricted their initial application to high-value artefacts. This in turn allowed their rapid perfection.

21  F.C. Copleston, *Aquinas*, 1991, pp. 63–69. The Catholic theologian was bound to use the pagan philosopher's thinking as a vehicle for innovation. The resistance to receiving historical Aristotelianism promoted this. 'The Greek philosopher was concerned with the problem of "motion" in the wider sense of becoming, whereas Aquinas made the problem of existence the primary metaphysical problem.'

22  Crowther, *Francis Bacon: The First Statesman of Science*, 1960. Bacon's combination of politics with technological determinism appealed to theorists of communism.

23  Angelo Maggi and Nicola Navone, *John Soane and the Wooden Bridges of Switzerland*, 2003. Palladio's recreation of Caesar's bridge in *Quattro Libri* was preceded by a series of detail studies of joints and construction techniques.

24  Plato, *Meno*. In this Socratic dialogue the concept generates the sophistic paradox: If you don't know it already how can you recognise it. If you do know already then what is being strived for?

25  Heyman, *The Stone Skeleton*, 1997. Bypassing construction rationales in favour of pure forms released structural engineering to the study of abstractions. The dome is reconceived as a membrane.

26  Kurrer, *Geschichte der Bauschichte*, 2002, pp. 121–124. Forms assembled from the platonic solids read well. The curves obtained have no special structural significance but occasionally approximate the structural ideal.

27  Morrissey, 2005. The Baroque is a rich amalgam from which to draw personality traits and emotional engagements.

## Chapter 6

1  Richard Dawkins, 2006. Memes are any unit of cultural information. Quite how they are reduced is obscure. The technology of building arches is cited as an example.

2  Baroque dismemberment. The sectioning of architectural elements at the same time as of the body in anatomical studies.

3  Norman Hampson, *The Enlightenment*, 1990. Different starting points generate different enlightenments.

4  Pocock, *Barbarism and Religion*, 1990. The existence of separate enlightenments running in parallel is a startling discovery. That they persisted without coalescence is also intriguing.

5  Tinniswood, *His Invention So Fertile: A Life of Sir Christopher Wren*, 2002. A prosaic background is credited with allowing Wren to 'float through' the upheavals of his time. His character may have helped too.

6  Jardine, *On a Grander Scale*, 2002. Wren seemed able to assimilate influences with great fluency. Dutch construction and detailing informed his work during the Glorious Revolution.

7  Ackroyd, 1993. Ian Sinclair's psychogeography is applied to the curiosity of English Baroque.

8  Jardine, *The Curious Life of Robert Hooke*, 2004. The founding members of the Royal Society were conscientious in defining its values and limitations, sometimes to the detriment of it operation.

9   Deleuze, *The Fold: Leibnitz and the Baroque*, 1993. The work of the seventeenth-century philosopher is presented as a sensibility and complete system applicable to contemporary notions of the 'nomadic' nature of the subject with ideas migrating across pleats of a continuum.

10  Denis Diderot. He wanted all people to have access to all the knowledge in the world.

11  Bosse, *Manual On Perspective*, 1648. Engravings made with exact perspective were intended to convey information without any possibility of confusion or error.

12  Luigi Ficacci, 2000, Piranesi's chosen medium (etching over engraving) introduces effects to the detriment of the delineation. 'Arcadian' embellishments buffer the reality of his 'verduti'.

13  Casanova, *The History of My Life*, 1997, vol. I, trans. William R. Trask, ch. 8, p. 220. The Augustan replacement bridge at Narnia, Umbria, on the Via Flaminia eastwards from Rome had the biggest span of its age; a 32-metre-wide arch over the river Neva. Casanova wastes no ink on describing any of the ruin's features, engineering or otherwise.

14  Maggi and Navone, *John Soane and the Wooden Bridges of Switzerland*, 2003, pp. 31–55. Their designs were objects of enquiry during their lifetime and this interest encouraged their tireless innovation.

15  Peters, *Transitions in Engineering*, 1987, pp. 51–53. The combination in Coulomb's expositions of thought experiment failure diagrams, geometric analysis of movement, graphical statics and algebraic formulation to the minimum level required for a usable solution is totally modern.

16  Billington, *Robert Maillart's Bridges*, 1976. The inference is that the didactic function of the teacher can in practice be transmuted into a clarity of design intent.

17  Dava Sobel, 2005. The emphasis is on the importance of human will overcoming technical, social and economic barriers.

18  Baroque. Definitions are legion, many so dissonant as to make the term self-referential. The opposition of light and dark is important.

19  John Stoye, *The Siege of Vienna*, 2001, pp. 36–40. Italy, continually churning with warring city states and factions, provided the best engineers. Shortages of money and space together with the steady improvement of artillery made innovation essential.

20  Cardwell, *The Fontana History of Technology*, 1994, pp. 113–128. The precocious development of the steam engine appears as a leapfrogging of practice and theory. Each advance prompted the direction of the next.

21  Giedion, *Space Time and Architecture*, 1974, pp. 169–210. Without realising it, Giedion compares the unmannered way iron was used by engineers with the architect's self-conscious uncertainty about how to use the new material.

22  Caisse Nationale des Monuments, *Soufflot et Son Temps*. Stone was being cramped together with iron to create the new spans and spaces. It still feels an uncomfortable thing to do.

23  Heyman, 'The Crossing Piers of the Pantheon', *Structural Engineer*, 8 August 1985. How the principal elements of structure may have been compromised by poor detailing.

24  Society for Protection of Ancient Buildings. The SPAB was founded in 1877 by the theorist William Morris. France's Caisse Nationale des Monuments Historique appeared only in 1914. Attempts to unravel shady preservation and misleading research from earlier on need their own techniques of recovery.

25  J.E. Gordon, *The New Science of Strong Materials or Why You Don't Fall Through the Floor*, 1976. It is wonderful that material characteristics can effect such a step change of performance between the world's environments and that humans can exploit these idiosyncrasies to their advantage.

## Chapter 7

1   Zamoyski, *1812: Napoleon's Fatal March on Moscow*, 2005. The tactic of using allies to bear the brunt of the fighting while one's own elite completes the flanking manoeuvre is as old as the Roman Empire. It was Napoleon's most effective ploy, preserving his best troops for another day. The practice was not unknown to the British army in the First World War.

2   Said, 2003, pp. 80–88. Napoleon's academic entourage fronted his contacts with the occupied peoples, all the time presenting the fighting as being on behalf of Islam.

3  Booker, *A History of Technical Drawing*, 1963, pp. 86–106. Monge's contribution is set alongside independent work, from Britain and America, particularly that of William Farish. The latter was formulated specifically to meet mechanical engineering problems.

4  Kranakis, *Constructing a Bridge*, 1997, pp. 157–164. The notion of 'social shaping' is introduced to explain Navier's total reliance on theory.

5  John Anderson, *A History of Aerodynamics*, 1999, pp. 89–91. Navier's notorious setback is contrasted with his little-known yet lasting contribution to fluid dynamics.

6  Hawking, *God Created the Integers*, 2005. Persuaded by Lagrange and Laplace to abandon engineering in favour of mathematics, Cauchy was a staunch royalist and religious bigot, unpopular in Republican times.

7  Jean-Baptiste Rondelet, *Traite Theoretique et Practique del'Art de Batir*, 1802. The taxonomy is based on a mixture of structural systems (arches, buttresses, etc.), building elements (capitals, bases, etc.) and architectural units (colonnades, doors, domes).

8  Christopher Alexander, *A Pattern Language*, 1977. An ambitious study makes all architectural design into the combination of a finite number of patterns.

9  Dorinda Outram, *Georges Cuvier: Vocation, Science and Authority in Post-Revolutionary France*, 1984. In founding palaeontology and deploying comparative methods of biology, Cuvier was working to replace creationism with a new rationale.

10  D'Arcy Thompson, *On Growth and Form*, 1992. The Victorian biologist initiated mathematical descriptions of biological transforms, including structural influences on organic form.

11  Nicholas Pevsner, *Ruskin and Viollet-Le-Duc: Englishness and Frenchness in the Appreciation of Gothic Architecture*, 1969. Viollet-le-Duc's rationalism is contrasted with Ruskin's emotion in their encounter with Gothic architecture. In their works of criticism, both men were seeking a drive for a contemporary architecture.

12  Telford: The Father of Modern Engineering, 2007. An exhibition to mark his 250th anniversary places Telford as the progenitor of civil engineering being art.

13  John Rapley, *The Britannia and other Tubular Bridges*, 2003, p. 113. The obsolescence of a structural form is seldom traced so well. The very rapid demise of tubular construction makes it a good case study.

14  Cardwell, *Fontana History of Technology*, 1994, p. 260. Son of a village shoemaker, Richard Roberts moved to Manchester and created a wide range of machines and products (including cast iron billiard tables for the termite-infested colonies). His prolific invention culminated in first prize at the Great Exhibition of 1851 for a clock.

15  Vaughan, *Isambard Kingdom Brunel: Engineering Knight Errant*, 1991, pp. 207–217. Ever the showman, Brunel appropriated Stephenson's method of lifting tubes into place, performing the initial floatation before huge crowds and directing operations from a dais with coloured flags.

16  L.T.C. Rolt, 2000. The ethics-free interpretation of Victorian engineering achievement is fortunately over.

17  Christopher Silver, *Renkioi: Brunel's Forgotten Crimean War Hospital*, 2007. The influence of this example, complete integration of function and engineering systems all designed from first principles, may have been curtailed by association with the Crimean setback to empire.

18  Rapley, *Thomas Bouch: The Builder of the Tay Bridge*, 2008. Ambition was one of his traits, and Bouch died a recluse two years after the disaster.

19  Kranakis, *Constructing a Bridge*, 1997, pp. 280–287. An 'American engineering tradition' is identified, though oddly not coupled to a Scottish tradition. An active distrust of theory is one of its traits.

20  David MacGregor, *British and American Clippers*, 1993, pp. 52–87. A close analysis of hull lines reveals how these ships were honed for speed.

21  Basil Lubbock, *The Nitrate Clippers*, 1976. The Germans were importing guano from Chile in huge quantities – a source of nitrate to make artificial fertiliser and explosives, the foundation of their chemical industry.

22  Addis, *The Art of the Structural Engineer*, 1994. The resilient wrought iron framing and light strong sheathing of elm and teak are ideal structural dispositions. The maintenance of the vessel is a nightmare of electrolytic corrosion and timber deterioration.

23  Kenneth Edwards et al., *The Four-masted Barque 'Lawhill'*, 1998. A detailed description of a technological system at the peak of perfection resisting its decline. The appropriation of a new construction technique turns the original form into something else unique.

## Chapter 8

1  Hobsbawm, *The Age of Capital, 1848–1875*, 1996, pp. 234–236. This analysis of the struggle between north and south also identifies incompatible attitudes to technology, based on urban and agrarian sensibilities respectively, fuelling the hatred.

2  A true designer of manufactured artefacts and machines. Whitney's most famous contribution was the cotton gin.

3  Before 1850 guns had been hand-finished, with components eased together. The Springfield carbine used breech blocks machined to a close enough tolerance for any bolt to fit any rifle. 'Armory practice' and Marc Brunel's compartmentation of the manufacturing processes were combined by the car manufacturer Henry Ford to become the modern production line.

4  Trains would cross the bridge driving a wave of distortion ahead of themselves. The need for stiffness in the deck and flexibility in the superstructure was made manifest. Tie-downs improved the proportion of immutable, stabilising 'dead' load to changeable, disturbing 'live' load.

5  The caisson followed immediately on experiments with diving bells. Engineers used air pockets trapped beneath heavy containers to inspect submerged sites. Salvage experts and treasure seekers began to delve beneath the ooze.

6  Plowden, 1974. The need not only to improve upon but also to differentiate one's invention from the competition appears to have encouraged variation.

7  The Bollman truss was used exclusively by the Baltimore and Ohio railway. Railway companies' proprietary rights to the various structural systems exploited patents for indirect commercial advantage.

8  Pacey, *The Maze of Ingenuity*, 1992. As well as technical responses to the failures, regulatory controls, hampered by the separation of state and federal powers, were constituted and imposed.

9  Bollman delegated the analysis of his bridges to Fink.

10  Arthur E. Morgan, *Dams and Other Disasters*, 1971, pp. 91–125. Eads's experiences at the hands of the US Corps of Engineers provides a text book example of the behaviour of a technical establishment towards innovative individuals.

11  Egyptian place names along the Mississippi river (Cairo, Memphis and Thebes) seem to allude to similarities with the river Nile and the American landscape's function as a granary.

12  Pacey, *The Maze of Ingenuity*, 1992. A talented amateur, directing but dissociated, together with established craftsmen seems to be a recurrently successful combination for design.

13  The wreck of the *Huntley* was discovered off Sullivan's Island, Charleston, in May 1995 by Clive Cussler, author of best-selling adventures.

14  Giedion, 1974. The development of the North American skyscraper is portrayed as a text-book example of technical determinism.

15  Giedion, 1974. Other important contributors include George Root, Louis Sullivan and James Hood

16  Henri Loyrette, *Gustave Eiffel*, 1985, p. 100. The entire design was analysed and detailed by four men.

17  De Lesseps's venture failed because malaria and yellow fever could not be controlled among the workers toiling through the Panamanian jungle.

18  Graefe, 1990. The 'gittermasts', hyperboloids formed of light metal grids, were standard components, simply jointed with an ingenious method of construction.

19  Vladimir Nabokov included Tolstoy in his seminar on modernism held at Cornell University. Anna Karenina goes under a train.

20  Kurrer, *Geschicht der Baustatik*, 2002, pp. 294–309. Strength, minimal deflection and light weight were all objectives important to crane design. High-strength materials and sophisticated manufacturing departments combined to offer crane designers unprecedented opportunities to refine structures.

21 Billington, 2003. As mentors to a coming generation, Culmann and Ritter provided method (graphical statics) and milieu (design consciousness) in their institution.

22 W.S. Hemp, *Optimum Structures*, 1975, pp. 71–101. The conditions for finding an optimum structure are set out. An absolute ideal may be impossible to achieve.

23 Leonard Eaton, *Hardy Cross and the Moment Distribution Method*, 2001. Hardy-Cross's reputation as a great teacher preceded his invention of the moment distribution method. The dean of his faculty was jealous and tried to remove him for insufficient publishing. He eventually filed a ten-page paper entitled 'Analysis of Continuous Frames by Distributing Fixed-End Moments'. The method was immediately adopted by practising engineers.

24 Science fiction is a deep-rooted genre tracing itself back to the romantic novels of the early nineteenth century, typically Mary Shelley's *Frankenstein*.

25 Jules Verne, *Twenty Thousand Leagues Under the Sea*, 1870, Norton Critical Edition. Verne's reading can be precisely traced from his technical descriptions. His research was accurately transcribed without speculation.

## Chapter 9

1 Newby, *Early Reinforced Concrete*, 2001, Studies in the History of Civil Engineering, 11. This collection of papers reflects the diversity of influences on the early development of reinforced concrete construction and the extraordinary prototypes it produced.

2 John MacKenzie, *Orientalism*, 1995, p. 77. Exotic planting and landscaping contributed to Victorian Orientalism but also proved the limits of its influence on western urbanism.

3 Gustav Flaubert, 1964. The author creates a modern style of writing to depict modern sensibilities.

4 Kurrer, *Geschichte der Baustatik*, 2002, pp. 354–358. Hennebique's clear and simple diagrams eased his system's reception among engineers.

5 Sarah Gaventa, *Concrete Design* (2006). The unusualness of concrete has prompted many designers to exploit its properties in extreme ways, from domestic implements to industrial monuments.

6 Picon, *L'Art de l'Ingénieur*, 1997. Paul Cottancin is identified as an engineer with an artistic sensibility beyond the conventional engineering discourse of his time.

7 Newby, 2001. Instead of a conceptual separation of concrete and steel, Cottancin envisaged the material as a homogeneous ductile substance. The even distribution of reinforcement that resulted from this approach now once again finds favour among engineers.

8 David Brown, *Bridges: Three Thousand Years of Defying Nature*, 2001, pp. 122–123. Low-quality cement in early reinforced concrete projects would have aggravated the shrinkage problem. Freysinnet's designs always made much of the problem.

9 Christopher Dean, *Housing the Airship* (1989). The extremely expensive aircraft were usually counterpointed by crude enclosures of trussed steel framing.

10 Giedion, *Space Time and Architecture*, 1952, pp. 328–331. Perret is placed as progenitor of the reinforced concrete frame unadorned.

11 Billington, 1976, pp. 33–38. Maillart's student notes record the transfer of Wilhelm Ritter's ideas on bridge aesthetics and the use of graphical statics to generate form.

12 World Heritage Citation, No. 1217. Viscaya Bridge. The first transporter bridge is astonishingly attenuated: a completely confident design. Its rigged construction enabled its economic reinstatement after being dynamited on the last day of the Spanish Civil War.

13 B.R. Mawson and R.J. Lark, 'Newport Transporter Bridge', *Structural Engineer*, vol. 77 (1999), no.16, pp. 15–21. The niche nature of the system led to the rapid 'roll-out' of a single design with minor progressive improvements across successive projects.

14 Plowden, 1974. Lindenthal's switch towards mass away from filigree is well illustrated and unexplained. He progressively raised live load allowances far above realistic levels.

15 Plowden, 1974, pp. 172–175. Quebec Bridge. The image of a grumpy old man, Theodore Cooper, refusing to come to his hotel room door when disturbed by the resident engineer Norman McClure, panicked and seeking guidance encapsulates the whole tale.

16  Ferguson, *Engineering and the Mind's Eye*, 1992, pp. 20, 24 and 25. It is suggested that Strauss's weird first pass at the design for the Golden Gate Crossing originated in his visual memory rather than the desire to remain in a comfort zone of experience.

17  Darl Rastorfer, *Six Bridges* (2000), pp. 25–29. The antagonism between Ammann and his former boss Lindenthal did much to entrench their positions as exponents of heavy-duty and lightweight design respectively.

18  Richard Scott, *In the Wake of Tacoma*, 2001, p. 16. Small deflection theory was a design method invented by Joseph Melan (1853–1941) for reinforced concrete arches and applied to suspension bridges by Leon Moisseiff (1872–1943). The success of the latter's Manhattan bridge of 1909 led Ammann to devote much time to the theory's development.

19  Arthur Morgan, *Dams and Other Disasters* (1971). A close analysis of one of the pathologies that resulted from North American 'big engineering'.

20  Kurrer, *Geschichte der Baustatik* (2002), ch. 8. Reinforced concrete is presented as having influenced the entire direction of structural engineering's development.

21  Pippard and Baker, *The Analysis of Engineering Structures* (1957). The book's structure, bolting more modern ideas onto a treatment of classical theory (analysis based on the calculus), is a fascinating embodiment of a historical process.

## Chapter 10

1  John David Anderson, *Introduction to Flight*, 2004, p. 18. Lilienthal recognised that for man to fly direct experience of control in the air was essential.

2  Arthur Barron, *The Wright Brothers*, TV programme, WNET, 1973. The creative tension that must have existed between two talented brothers is recreated using the creative tension that exists between two talented brothers.

3  Peter Wykeham, *Santos-Dumont* (1980). The emotional engagement of the aviation pioneers is well rendered.

4  Peter Joseph Capelotti, *By Airship to the North Pole* (1999). The polar expeditions by airship of 1907 and 1909 were hopeless: unplanned, romantic enactments of Victorian science fiction.

5  Tim Coates, *R101 The Airship Disaster* (1930). The official report of the crash investigation shows how technical failure is so often seeded by bureaucratic incompetence.

6  Richard K. Smith, *The Airships Akron and Macon* (1965). The niche technology, biplane scouts with no undercarriages, hooking onto 'trapeze', lasted just eleven years.

7  Christine Boyer, *Aviation and the Aerial View: Le Corbusier's Spatial Transformations in the 1930s and 1940s* (2003). The seriousness of these proposals shows how inconceivable the consequences of saturation bombing was at the time. Corb characteristically makes the technical proposition part of a general reorientation of space brought about by 'aviation'.

8  Jack Bruce, *The SE5A: Aircraft in Profile* (1965). The accretive process of early biplane design is well illustrated in the way the prototype's initially weak wing structure (it killed its designer) was readily modified to make the front-line planes capable of the strongest dive.

9  Heiner Emde and Carlo Demand, *Conquerors of the Air* (1968). The first transatlantic flight took the plane beyond its operational boundaries. The navigator, Arthur Brown, had to get out on the nose to scrape off the ice. The plane ditched in an Irish bog.

10  Derek James, 2001. The technical advances made in the competition found almost immediate applications in warplane design. A parallel discourse was set up adjacent to designs based on first war experience. Compare the Hawker Hurricane and Supermarine Spitfire.

11  Neville Shute, *Slide Rule* (1954). The engineering ambience in which the R100 was created is described, the privately funded alternative to the Air Ministry's R101, an aircraft capable of flying the 'Empire Routes' to India and transatlantic to Canada.

12  Robert Massie, *Dreadnought* (1992). The advent and technology of big gun ships is set within a very wide context of personalities and feelings.

13  Janusz Skulski, *The Heavy Cruiser Takao* (1994), p. 12. Reducing hull weight to outflank treaty limitations on displacement caused stability problems. The ships still needed to carry full

superstructures, and initially hull shapes were not modified sufficiently to compensate for raised centres of gravity.

14  Edwyn Gray, *Hitler's Battleships* (1999). The 'pocket battleships' could outgun anything that could catch them and outrun anything that could sink them. Their oil-fired boilers used fuel of better calorific value than coal, so better range than other cruisers. There was no tell-tale puff of smoke when they sighted a target and put on speed.

15  James Gordon, *The New Science of Strong Materials, or Why We Don't Fall Through the Floor* (1991). The localised stresses that occur as welds cool were initially a major problem in the assembly of liberty ships which might 'unzip' in high seas. With the success of asdic submarine detection, for a short while structural failure became a greater danger to merchant mariners than torpedoes. Of 4,700 freighters produced, 200 suffered catastrophic failure and another 1,200 severe damage.

16  Jacques Heyman, *Structural Analysis*, 1998, ch. 9, pp. 127–153. Plastic theory is presented as the pinnacle of structural theory's historical development.

17  John Milsom, *Russian Tanks, 1900–1970* (1975). The first five-year plan (1928) for the industrialisation of agriculture was important. The huge factories set up in the Urals and Siberia to mechanise farming were readily switched to tank production.

18  Fritz Kohl and Eberhard Rossler, *The Type XXI U-Boat* (1991), p. 8. The simple insight of a massive increase in battery compartments came to two junior engineers presented with an experimental design that required a huge fuel tank. A conventional configuration had become so well established that the visualisation of additional space had to be prompted from outside the accepted format.

19  James Gordon, *The Science of Structures and Materials* (1988), pp. 81–97. Fracture mechanics take engineering back into a real world of scale and specificity.

20  John McDonald, *Howard Hughes and the Spruce Goose* (1981). That critics could maintain that a plane of such size simply could never fly says much about the rhetoric that stays attached to technical matters.

21  Peter Bowers, *Boeing B52 A/H Stratofortress, Aircraft in Profile* (1973). The five-man team working Thursday–Sunday, 21–24 October 1948, in the Hotel van Cleve in Dayton, Ohio, is Boeing company legend. They were extending a long development series of designs using a variety of engines. The finished design required 3,000,000 man hours to complete.

22  Joe Sutter, *Creating the World's First Jumbo Jet and Other Adventures from a Life in Aviation* (2006). Take-off weights for the variants range between 330 and 440 tonnes.

23  Lancelot Law Whyte, *Aspects of Form* (1951). Gestalt theory suffered from its advocate's insistence on its universality.

24  Monk, *Ludwig Wittgenstein: The Duty of Genius* (1991), p. 26. To the question 'what is force?', Heinrich Hertz proposed the solution of restating Newtonian physics without it: 'When these painful contradictions are removed the question as to the nature of force will not have been answered; but our minds, no longer vexed, will cease to ask illegitimate questions.'

25  Timothy Clark, *Martin Heidegger*, 2001, pp. 9–25. The carpenter's feeling for the wood is presented as a supra-theoretical way of imposing form, techne rather than technology. The engineer must get beyond the superficial appearance of things in order to understand.

## Chapter 11

1  Fritz Stern, *Einstein's German World* (2001). Raymond Aron: 'The twentieth century could have been Germany's century.'

2  Felix Candela, *Hacia Una Nueva Filosophia De Las Estructuras* (1962). The pursuit of elegance permeates the work; soluble partial-differential equations yield beautiful simple edge profiles.

3  Guy Nordenson, *Seven Structural Engineers: The Felix Candela Lectures* (2008), p. 170. At the first congress of the International Association for Shell Structures, Isler presented a short paper entitled 'New Shapes for Shells'. His proposition that 'free' forms could be legitimately generated from analogues raised a storm of discussion.

# Notes

4   Frei Otto, *Finding Form: Towards and Architecture of the Minimal* (1995). The preoccupation with reducing weight coincided with the advances made in aeronautical engineering optimising air-frames.

5   Fritz Leonhardt, *Bridges: Aesthetics and Design* (1984). The aesthetic principles and taxonomy of bridges set down in this book seem rather simplistic and premature in view of what followed. They offer a way of thinking.

6   Le Havre, UNESCO World Heritage Citation Ref:1181. 'Le Havre is exceptional among many reconstructed cities for its unity and integrity.'

7   Anthony Alofsin, *Frank Lloyd Wright: The Lost Years, 1910–1922*, 1994. The adoption, development and eventual rejection of the 'textile block' system of construction by its author is analysed in the context of the experiment's wide influence on others.

8   Eduardo Torroja, *Philosophy of Structures* (1958). Beautiful illustrations, sketches of structure, support this contemplation of engineering's potential.

9   Giorgio Boaga and Benito Boni, *The Concrete Architecture of Riccardo Morandi* (1966). The attention of the architectural critic Kenneth Frampton has done much to situate Morandi's contribution.

10  Ada Louise Huxtable, *Pier Luigi Nervi*, Masters of World Architecture (1976). The descent into expressive, irrational form is well documented in the sequence of projects.

11  Bertrand Goldberg, 1975. His pre-occupation with construction and its processes led Goldberg into many experiments with unconventional forms out of mundane materials: cars, furniture and demountable accommodation.

12  A 'super-tall' tower over about forty storeys high where the shortening effects of self-weight during construction must be considered in the structural design.

13  Alvin Rosenbaum, *Works in Progress* (1994), pp. 182–183. Mega-frames always look their most impressive at the construction stage.

14  Yasmin Sabina Khan, *Engineering Architecture* (2004). A daughter's intellectual biography of her father, traces the development of a series of engineering ideas guided by the market and unencumbered by overweening philosophising.

15  Karl Terzaghi, *Theoretical Soil Mechanics* (1943). Based on a German text of 1925, here is an Austrian in the age of Freud trying to bring a scientific basis to his discourse.

16  Martin Pawley, 'Konrad Wachsmann: The Greatest Architect of the Twentieth Century'. *Architects' Journal*, 2 December 1999. The title's claim is based on Wachsmann's ambition to transform architecture through new means of construction. But how could a joiner simply ignore joint slippage?

17  Richard Buckminster Fuller, *Nine Chains to the Moon* (1938). The arch-text of twentieth-century technical idealism. The book established Buckminster Fuller's pre-eminence and his aftertaste of charlatanry.

18  Peter Jones, *Ove Arup* (2006), pp. 266–281. The 'key speech' is an important part of the Arup iconography. He did a lot to promote aphorism among engineers, as the book in hand shows.

19  Peter Murray, *The Saga of the Sydney Opera House* (2003). Bigger than just a building project, the creation of the opera house is a site on which to speculate on design, construction and culture in all their manifestations.

20  Nicholas Bullock, *Building the Post-war World* (2002). Brutalism was cheap; its uncommunicative abstraction suited post-war austerity. At Hunstanton School, the water tank featured prominently; the motif of expressed services carried over into the high-tech style after the direct juncture of elements had been replaced by over-fetishised sealed seams.

21  Henry Petroski, *Design Paradigms*, 1994, pp. 144–164. Failure is presented as an essential part of design progress. A 'design climate' must develop, in which events then take place.

22  Michael Light, *One Hundred Suns* (2003). 006 Trinity test, 16 July 1945

23  Norman Polmar and Kenneth Moore, *Cold War Submarines* (2003). 2,500 tonnes displacement for a patrol submarine at the end of the Second World War; 33,000 tonnes for the very biggest ballistic missile launchers today.

24  George Dyson, *Project Orion* (2002), pp. 270–273. The very characteristic that made the project so difficult in reality was to be the fictive initiative for alien action as a human spacecraft was detected 'throwing off a spray of radiation like the wake of a racing speedboat'.

25 The partial test ban treaty stopped the project.

26 Some 3,000 missiles were launched, killing approximately 7,250 people. Over 20,000 slaves died constructing them.

27 Harold Martin, *Introduction to Matrix Methods of Structural Analysis* (1966). The key text for modern structural engineers, written by a Professor of Aeronautics and Astronautics, Consultant to the Boeing Company.

28 Vidar Thomée, *From Finite Differences to Finite Elements Analysis* (1999). That there are still several incompatible histories of the development of finite element analysis competing for orthodoxy indicates its significance.

## Chapter 12

1 Peter Nicholls and John Clute, *The Encyclopaedia of Science Fiction* (1993). Science fiction is non-speculative and divides into genres. It shares these characteristics with design. It is also a genre of literature in general. Design is a genre of what?

2 Washington, Government Printing Office, 1947. The principles on which the huts were developed, simplicity of assembly, minimal shipping volume and weight and special accessories, ovens and beds were carried over into other product design after the war – all characteristics required to penetrate and build global markets. Compare also the trajectory of the Willis jeep or the plywood splints that became Eames's orthopaedic chairs for business use.

3 Susan Solomon, *Louis Kahn's Trenton Jewish Community Center* (2000), pp. 136–138. Servant and served spaces. An incredibly powerful concept was worked out in a tiny building before being exploited to its full potential across all types of building.

4 Brendan Gill, *Many Masks: A Life of Frank Lloyd Wright* (1998). Sending oneself a telegram congratulating oneself on a structural success may seem excessive but was only part of a general 'construction programme', drawings and theorising, to create a total architectural discourse above and beyond reality.

5 James Steele, *Eames House: Charles and Ray Eames* (2002). Two designers apply a real architectural sensibility to the organisation of discarded war products.

6 Peter Carter, *Mies Van Der Rohe at Work* (1974). The welded connections of the Farnsworth house and Barcelona chair, the build-up of the Seagram building mullions and Lakeshore Apartment corners required prodigious ingenuity to achieve their apparent simplicity. 'Simplicity is not simple' precludes irony.

7 Joseph Rose, *A Constructed View: The Architectural Photography of Julius Shulman* (2008). The celebrated picture makes the cover of the book.

8 Constructivism. A Russian movement, 1919 to 1934, promoting art and architecture's social purpose.

9 Team 4 disbanded in 1967 and two of its members went on to dominate contemporary British architecture: Norman Foster and Richard Rogers.

10 Kisho Kurokawa, *Metabolism in Architecture* (1977). The 'Metabolism Group' of young Japanese architects formed itself at the World Design Conference in 1960.

11 The 'P delta effect' is a term describing how the self-weight of a displaced shape tends to drag it over. There is an apparent decrease in lateral stiffness as distortions get bigger. In heavy buildings with light bracing, the phenomenon can add a 10 to 25 per cent increase in equivalent lateral load on the structure.

12 Cecil Balmond, *Informal* (2002). The complex will have to get a lot more complicated to match the chaos-ordering capability of a designer's mind.

13 George Bridgeman, *Life Drawing* (1924). A tutor at the Art Students League at New York, Bridgeman developed a systematic approach to reprising the figure drawing style of Michelangelo.

14 Kate Colquhoun, *A Thing in Disguise: The Visionary Life of Joseph Paxton* (2004), The subtitle, 'The busiest man in England', highlights that English reconciliation of the cults of effortlessness and busyness.

15  David Billington and David Billington Jr, *Power, Speed and Form* (2006). Having achieved a fully formed critical method specific to technology, the Billingtons apply it to a sample of modern artefacts.

16  The vessels had positive buoyancy, so the discovery of a wreck is unlikely.

17  Literature, depiction, the way things are made: all contribute in reconstructing a bygone intellectual milieu.

18  Sharpened steel pens produced graphics like engravings made for the 'machine-for-living-in'. Meanwhile, soft pencil rendered the deep chiaroscuro of the California sunshine to produce la Miniatura.

19  Adolf Loos, *Rules for Those Building in the Mountains* (1913). 'Build as well as you can. No better. Do not outstretch yourself. And no worse. Do not deliberately express yourself on a more base level than the one with which you were brought up and educated. This also applies when you go into the mountains. Speak with the locals in your own language. The Viennese lawyer that speaks to the locals in a country bumpkin's accent is beneath contempt. Pay attention to the forms in which the locals build. For they are the fruits of wisdom gleaned from the past. But look for the origin of the form. If technological advances made it possible to improve on the form, then always use this improvement . . . 'Be true! Nature only tolerates truth. It copes well with iron truss bridges, but rejects Gothic arched bridges with turrets and defensive slits.'

20  Mohsen Mostafavi, *Structure as Space. Engineering and Architecture in the Works of Jurg Conzett and his Partners* (2006). Two successive bridges across the Traversina gorge, one complicated the other simplicity itself, are separated by four years of design thought.

21  Ed van Hinte and Adriaan Beukers, *Lightness: The Inevitable Renaissance of Minimum Energy Structures* (1998). The independence of technical and economic arguments for lightness is revealed.

22  The French have resisted high-rise building. Since the vituperation heaped on the Eiffel tower, Parisians have always sought to maintain their skyline. A thirty-year-old ban on building above thirty-seven metres in the inner city is about to be lifted.

23  Ödön von Horvath, *Der Jüngste Tag*, 1936. A play about the impossibility of human accountability in the face of modern technology.

24  Eugene Kremer, *Re-examining The Citicorp Case: Ethical Paragon or Chimera* (2002). Focusing on a dramatic solution is a common way of drawing attention away from the original pattern of failings leading up to a problem.

25  Jann-Nan Yang and Fanis Giannopoulos, *Active Control and Stability of Cable-Stayed Bridges* (1989). A collaboration between a civil engineer and structural designer of automobiles demonstrating how feedback systems can improve structures.

# Bibliography

Ackroyd, P. (1993) *Hawksmoor*. London: Penguin.

Ackroyd, P. (1994) *The House of Doctor Dee*. London: Penguin.

Addis, W. (1994) *The Art of the Structural Engineer*. London: Ellipsis London Ltd.

Alexander, C. (1977) *A Pattern Language: Towns, Buildings, Construction*. Oxford: Oxford University Press.

Alofsin, A. (1994) *Frank Lloyd Wright—the Lost Years, 1910–1922: A Study of Influence*. Chicago: University Of Chicago Press.

Anderson, J. (2004) *Introduction to Flight*. McGraw Hill.

Anderson, J.D. (1999) *A History of Aerodynamics: And Its Impact on Flying Machines*. Cambridge: Cambridge University Press.

Archimedes (2002) *The Works of Archimedes* trans. Sir Thomas Heath. New York: Dover.

Balmond, C. et.al. (2002) *Informal*. New York: Prestel USA.

Baker, A. and Pippard, J.F. (1957) *The Analysis of Engineering Structures*. London: Edward Arnold.

Baker, G.P. (2001) *Constantine the Great*. Natl Book Network.

Ball, W. (2001) *Rome in the East*. Routledge.

Barral i Altet, X. (2002) *The Early Middle Ages: from Late Antiquity to A.D. 1000*. Cologne: Taschen.

Barron, A. (1973) *The Wright brothers* (TV programme) WNET.

Bathhurst, B. (1999) *The Lighthouse Stevensons*. London: Flamingo.

Beukers, A. and van Hinte, E. (1999) *Lightness: The Inevitable Renaissance of Minimum Energy Structures*. Rotterdam: Uitgeverij 010 Publishers.

Biagioli, M. (1993) *Galileo Courtier*. University of Chicago.

Billington, D. (1976) *Robert Maillart's Bridges*. Princeton.

Billington, D.P. (2003) *The Art of Structural Design: A Swiss Legacy*. Princeton, NJ: Princeton University Art Museum.

Billington, D.P. and Billington, D.P. Jr. (2006) *Power, Speed, and Form: Engineers and the Making of the Twentieth Century*. Princeton, NJ: Princeton University Press.

Boaga, G. and Boaga, B.B. (1966) *The Concrete Architecture of Riccardo Morandi*. Connecticut: Praeger.

Bonner, A. (1993) *Doctor Illuminatus: A Ramon Llull Reader*. Princeton.

Booker, P.J. (1963) *A History of Engineering Drawing*. London: Chatto & Windus.

Bowers, P. M. (1973) *"Boeing B-52A/H Stratofortress"*. Aircraft in Profile, Volume 13. Berkshire: Profile Publications Ltd.

Boyer, C. (1989) *History of Mathematics*. Wiley.

Boyer, C.B. (1949) *The History of the Calculus and its Conceptual Development*. New York: Dover.

Boyer, M.C. (2003) Aviation and the Aerial View: Le Corbusier's Spatial Transformations in the 1930s and 1940s. *Diacritics* 33 (3/4) Fall–Winter 2003 93–116.

Braudel, F. (1996) *The Mediterranean*. University of California.

Braudel, F. (2001) *The Mediterranean in the Ancient World*. London: Penguin.

Bridgeman, G. (1924) *Bridgeman's Life Drawing*. New York: Dover Publications.

Brown, D. (2001) *Bridges: Three Thousand Years of Defying Nature*. MBI.

Brown, D.J. (1993) *Bridges: Three Thousand Years of Defying Nature*. London: Mitchell Beazley.

Brown, P. (2000) *Augustine of Hippo*. University of California.

Bruce, J.M. (1965) *"The S.E.5A"*. Aircraft in Profile, Volume 1/Part 1. Windsor: Profile Publications Ltd.

# Bibliography

Bucher, F. (1960) 'Cistercian Architectural Purism' *Comparative Studies in Society and History* 3 (1) October: 89–105.

Buckminster-Fuller, R. (1936) *Nine Chains to the Moon*. Southern Illinois University.

Bullock, N. (2002) *Building the Post-War World*. London: Routledge

Butterworth A. and Laurence R. (2006) *Pompeii: The Living City*. New York: St Martin's Press.

Byron, R. (2001) *The Road to Oxiana*. London: Penguin.

Caisse National des Monuments. (1980) *Soufflot et son Temps*. CNMHS.

Cameron, A. (1996) *Procopius and the Sixth Century*. London: Routledge.

Candela, F. (1962) *Hacia Una Nueva Filosofia De Las Estructuras* Buenos Aires: Ediciones 3.

Capelotti, P.J. (1999) *By Airship to the North Pole: An Archaeology of Human Exploration*. New Brunswick, NJ: Rutgers University Press.

Cardwell, D.S.L. (1994) *The Fontana History of Technology*. Glasgow: Fontana Press.

Carter, P. (1974) *Mies Van Der Rohe at Work*. Connecticut: Praeger.

Casanova, G. (1997) *History of My Life. Volume 1,* trans. William R. Trask. Baltimore: John Hopkins University Press.

Casson, L. (1959) *The Ancient Mariners: Seafarers and Sea Fighters of the Mediterranean in Ancient Times*. New York: Macmillan.

Casson, L. (1971) *Ships and Seamanship in the Ancient World*. Baltimore: John Hopkins University Press.

Chandrasekhar, S. (1994) *Newton and Michaelangelo*. Lecture at 44th Meeting of Nobel Laureates, Lindau, 28 June 1994.

Chavaloux, R. (2001) *Nombre d'Or, Nature et Oeuvre Humaine*. Chalagram.

Chen, W.F. and Scawthorn, C. eds. (2003) *Earthquake Engineering Handbook*. Florida: CRC Press.

Chilton, J. (2000) *Heinz Isler: The Engineer's Contribution to Contemporary Architecture*. London: Thomas Telford.

Christiansen, E. (2002) *The Norsemen in the Viking Age*. Wiley.

Clark, T. (2001) *Routledge Critical Thinkers: Martin Heidegger*. London: Routledge.

Coates, T. ed. (1999) *R101: The Airship Disaster, 1930*. London: UK Stationery Office.

Colquhoun, K. (2004) *A Thing in Disguise: The Visionary Life of Joseph Paxton*. London: Harper Perennial.

Conrad, J. (1979) *The Nigger of the Narcissus*, ed. Robert Kimbrough. New York: W.W.Norton.

Copleston, F. (1991) *Aquinas*. Penguin.

Coulton, J. (1977) *Ancient Greek Architects at Work*. Cornell.

Courtenay, L.T. (1997) *The Engineering of Medieval Cathedrals. (Studies in the History of Civil Engineering, Vol.1)*. Aldershot: Ashgate Publishing.

Cresswell, K.A.C. (1969) *Early Muslim Architecture*. Oxford: Oxford University Press.

Crowther, J.G. (1960) *Francis Bacon: The First Statesman of Science*. London: Cresset Press.

Cuomo, S. (2001) *Ancient Mathematics*. Routledge.

Davis, L. (2003) *The Jupiter Myth*. London: Arrow.

Dawkins, R. (2006) *The Selfish Gene*. Oxford: Oxford University Press.

Dawson, S. (2003) 'Working details. Dublin's design pinnacle', *Architects' Journal*, (217)19, (15 May), 40–43.

Dean, C. (1989) *Housing the Airship*. London: Architectural Association.

Deleuze, G. (1993) *The Fold: Leibniz and the Baroque*, trans. Tom Conley. London: Athlone.

Deneux, H. (1927) *L'Evolution des Charpentes du XIe au XVe Siecle*. MONUM.

Department of the Navy (1947) *Building the Navy's Bases in World War II: History of the Bureau of Yards and Docks and the Civil Engineer Corps, 1940–1946. Vol. 1*. Washington, DC: US Government Printing Office.

Derek N. J. (2001) *The Schneider Trophy Contest: 1913–1931*. Gloucestershire: NPI Media Group.

Devreese, J. T. and Vanden Berghe, G. (2007) *Magic is No Magic: The Wonderful World of Simon Stevin*. Southampton: WIT Press.

Dietrich, R.J. (1998) *Faszination Brücken*. Munich: Callwey.

Ding, D. (1994) Ancient and Modern Chinese Bridges. *Structural Engineering International*, 4.1.

Dunkelberg, K. (1985) *IL31 Bamboo as a Building Material*. Stuttgart: Institute for Lightweight Structures.

Dyson, G. (2002) *Project Orion: The Atomic Spaceship 1957–1965*. London: E-Allen Lane.

Eaton, L. (2001) *Hardy Cross and the Moment Distribution Method*. Birkhäuser.

Eaton, L.K. (2001) Hardy Cross and the 'Moment Distribution Method' *Nexus Network Journal*, 3(3) Summer 2001 http://www.nexusjournal.com/Eaton.html.

Edwards, K. and Anderson, R. (1996) *The Four-masted Barque "Lawhill" (Anatomy of the Ship Series)*. London: Conway Maritime Press.

Emde, H. and Demand, C. (1968) *Conquerors of the Air: The Evolution of Aircraft 1903–1945*. London: Random House.

Evans, J.A.S. (2000) *The Age of Justinian: The Circumstances of Imperial Power*. London: Routledge.

Faegre, T. (1979) *Tents: Architecture of the Nomads*. New York: Anchor Press/Doubleday.

Ferguson, E.S. (1992) *Engineering and the Mind's Eye*. Cambridge, MA: The MIT Press.

Ferguson, N. (1997) *Virtual History: Alternatives and Counterfactuals*. London: Picador.

Ficacci, L. (2000) *Piranesi: The Complete Etchings*. Taschen.

Fifer, J.V. (1996) *The Master Builders: Structures of Empire in the New World-Spanish Initiative and United States Invention*. Edinburgh: Durham Academic Press.

Flaubert, G. (1964) *Sentimental Education* trans. Robert Baldick. London: Penguin.

Fletcher, B. (1948) *A History of Architecture on the Comparative Method for Students*. Craftsmen & Amateurs New York: Charles Scribner's Sons.

Fletcher, R. (1992) *Moorish Spain*. Wiedenfeld & Nicholson.

Forte, A. et al. (2005) *Viking Empires*. Cambridge.

Fuller, R. Buckminster. (1938) *Nine Chains to the Moon*. Philadelphia: Lippincott.

Galinsky, K. (1998) *Augustan Culture: An Interpretive Introduction*. Princeton: Princeton University Press.

Gallet, Michel A.O (1980) *Soufflot et son temps: 1780–1980 (Exhibition catalogue)*. Paris: Caisse Nationale des Monuments Historiques et des Sites.

Gardiner, R. and Greenhill, B. (1993) *Sail's Last Century: The Merchant Sailing Ship 1830–1930 (Conway's History of the Ship)*. London: Conway Maritime Press.

Gardiner, R. and Morrison, J. (1995) *The Age of the Galley: Mediterranean Oared Vessels Since Pre-classical Times (Conway's History of the Ship)*. London: Conway Maritime Press.

Garreta, A.A. (2002) *Skyscrapers*. Grabasa, Spain: Atrium Group.

Gaventa, S. (2006) *Concrete Design*. London: Mitchell Beazley.

Gernet, J. (1996) *A History of Chinese Civilisation*. Cambridge.

Gibbs, M. (2003) 'Eco tower: glass towers are not necessarily eco friendly. This one shows how new architecture can be evolved from ecological and humane principles', *Architectural Review*, August, 58–63.

Giedion, S. (1952) *Architecture and the Phenomenon of Transition. The Three Space Conceptions in Architecture*. Massachusetts: Harvard University Press.

Giedion, S. (1974) *Space, Time and Architecture: The Growth of a New Tradition (Charles Eliot Norton Lectures)*. Massachusetts: Harvard University Press.

Gil, V. (2003) *Guia arqueologica de Cordoba: una vision de Cordoba en el tiempo a traves de su patrimonio arqueologico*. Cordoba: Plurabelle.

Gill, B. (1998) *Many Masks: A Life Of Frank Lloyd Wright*. New York: Da Capo Press.

Gimpel, J. (1983) *The Cathedral Builders*. London: Grove Press.

Goddard, D. (1968) *Encyclopaedia of World Art*. McGraw Hill.

Goepfert, Yvette ed. (1982) *Pont du Gard*. Le Cannet: Editions Aio.

Goldberg, B. (1975) *Bertrand Goldberg (A+U)*. Tokyo: A + U Pub.

Golding, W. (1974) *The Spire*. Faber & Faber.

Goldsworthy, A. (2001) *The Punic Wars*. Cassell.

Gordon, J.E. (1988) *The Science of Structures and Materials*. New York: Scientific American Library.

Gordon, J.E. (1991) *Structures or Why We Don't Fall Through the Floor*. London: Penguin.

Gordon, J.E. (1991) *The New Science of Strong Materials, or Why You Don't Fall Through the Floor*. London: Penguin.

Gould, S. (1999) *Leonardo's Mountain of Clams and the Diet of Worms*. Three Rivers Press.

Gould, S.J. (1989) *Wonderful Life: The Burgess Shale and the Nature of History*. New York: W.W. Norton.

## Bibliography

Graefe, R. and Gappoev, M. (1990) *Vladimir G. Suchov 1853–1939 Die Kunst der Sparsamen Konstruction*, Stuttgart: Deutsche Verlags-Anstalt.

Gray, E.A. (1999) *Hitler's Battleships*. Annapolis, Maryland: US Naval Institute Press.

Greenhill, B. (1996) *The Evolution of the Sailing Ship*. Conway.

Greenhill, B. and Hackman, J. (1991) *Herzogin Cecilie: The Life and Times of a Four-masted Barque*. London: Conway Maritime Press.

Gutas, D. (1998) *Greek Thought, Arabic Culture: The Graeco-Arabic Translation Movement in Baghdad and Early 'Abbasid Society (2nd–4th/8th–10th centuries) (Arabic Thought & Culture)*. London: Routledge.

Hall, M. B. [1962] *The Scientific Renaissance, 1450–1630*. New York: Dover Publications.

Hampson, N. (1990) *The Enlightenment*. London: Penguin.

Harris, R. (1993) *Discovering Timber-Framed Buildings*. Shire.

Hasler, H. and McLeod, J. (1989) *Practical Junk Rig*. Tiller.

Hawking, S. (2005) *God Created the Integers: The Mathematical Breakthroughs That Changed History*. Philadelphia: Running Press Book Publishers.

Heath, T. (2002) *The Works of Archimedes*. Dover.

Hemp, W. S. (1973) *Optimum Structures*. Gloucestershire: Clarendon Press.

Henderson, G. (1968) *Chartres*. Harmondsworth: Penguin.

Hewitt, C. (1997) *English Historic Carpentry*. Linden

Heyman, J. (1985) 'The Crossing Piers of the Pantheon', *Structural Engineer*, 3 August.

Heyman, J. (1997) *The Stone Skeleton*. Cambridge: Cambridge University Press.

Heyman, J. (1998) *Structural Analysis: A Historical Approach*. Cambridge: Cambridge University Press.

Heyman, J. (1999) *The Science of Structural Engineering*. London: Imperial College Press.

Hill, D. (1984) *A History of Engineering in Classical and Medieval Times*. London: Routledge.

Hinte, E. van and Beukers, A. (1998) *Lightness*. Rotterdam: 010 Uitgeverij.

Hobsbawm, E.J. (1996) *The Age of Capital 1848–1875*. London: Vintage.

Hollingdale, S. (1994) *Makers of Mathematics*. London: Penguin.

Horden, P. and Purcell, N. (2000) *The Corrupting Sea: A Study of Mediterranean History*. Oxford: Blackwell Publishing.

Horvath, O. von and Krischke, T. (2001) *Der jüngste Tag und andere Stücke*. Frankfurt am Main: Suhrkamp.

Humphrey, J.W. and Oleson, J.P. (1998) *Greek and Roman Technology: A Source Book: Annotated Translations of Greek and Roman Texts and Documents*. London: Routledge.

Huxtable, A. L. (1976) *Pier Luigi Nervi (Masters of World Architecture)*. New York: George Braziller.

Inoue, A. (2003) 'Taipei International Finance Centre, Taiwan', *The Structural Engineer*, 81(8), (15 April), 15–17.

James, D. (2001) *The Schneider Trophy Contest 1913–1931*. NPI.

Jardine, L. (2002) *On a Grander Scale: The Outstanding Career of Sir Christopher Wren*. London: Harper Collins.

Jardine, J. (2004) *The Curious Life of Robert Hooke: The Man Who Measured*. London London: Harper Collins.

Jones, D. (1995) *Boats*. Egyptian Bookshelf.

Jones, P. (2006) *Ove Arup*. Yale.

Kenny, A. (2004) *Ancient Philosophy*. Oxford.

Khan, Y.S. (2004) *Engineering Architecture: The Vision of Fazlur R. Khan*. New York: W.W. Norton.

Khoury, N. (1996) *The Meaning of the Great Mosque of Cordoba in the Tenth Century*. Murganas 13.

King, R. (2001) *Brunelleschi's Dome: The Story of the Great Cathedral in Florence*. London: Pimlico.

Kohl, F. and Rossler, E. (1991) *The Type XXI U-boat (Anatomy of the Ship Series.)*. London: Conway Maritime Press.

Kranakis, E. (1997) *Constructing a Bridge: An Exploration of Engineering Culture, Design, and Research in Nineteenth-century France and America*. Cambridge, MA: The MIT Press.

Kremer, E. (2002) "(Re)examining the Citicorp Case: Ethical Paragon or Chimera. Paper at Ethics and Architecture" conference, New York, April 6, 2002, http://www.crosscurrents.org/kremer2002.htm.

Kunzle, D. (2002) *From Criminal to Courtier: The Soldier in Netherlandish Art 1550–1672*. Leiden: Brill Academic Publishers.

Kurokawa, K. (1977) *Metabolism in Architecture*. London: Studio Vista.

Kurrer, K.-E. (2002) *Geschichte der Baustatik (The History of Structural Engineering)*. Berlin: Ernst & Sohn.

Lane, F.C. (1934) *Venetian Ships and Shipbuilders of the Renaissance*. Baltimore: John Hopkins University.

Lark, R.J., Mawson, B.R. and Smith, A.K. (1999) The Refurbishment of Newport Transporter Bridge, *The Structural Engineer* 77 (16) 17 August 1999, 15–21.

Lattis, J. (1994) *Between Copernicus and Galileo*. University of Chicago.

Lavery, B. (2004) *The Ship*. DK Adult

Lawrence, C. (1984) *Medieval Monasticism*. Longman.

Leonhardt, F. (1984) *Bridges: Aesthetics and Design*. Cambridge, MA: MIT Press.

Levi, A. (2002) *Renaissance and Reformation: The Intellectual Genesis*. Connecticut: Yale University Press.

Light, M. (2003) *100 Suns*. New York: Knopf.

Lindberg, D.C. (1992) *The Beginnings of Western Science: The European Scientific Tradition in Philosophical, Religious and Institutional Context, 600 B.C. to A.D. 1450*. Chicago: University of Chicago Press.

Loos, A. (2002) *On Architecture*. Ariadne.

Loyrette, H. (1986) *Gustave Eiffel*. Paris: Payot.

Lubbock, B. (1976) *The Nitrate Clippers*. Glasgow: Brown, Son & Ferguson.

McDonald, J.J. (1981) *Howard Hughes and the Spruce Goose (Modern aviation series)*. Pennsylvania: TAB Books.

Macfarlane, A. and Martin, G. (2002) *The Glass Bathyscaphe: How Glass Changed The World*. London: Profile Books.

MacGregor, D.R. (1993) *British and American Clippers: A Comparison of their Design, Construction and Performance in the 1850's*. London: Conway Maritime Press.

MacKenzie, J. (1995) *Orientalism: History, Theory and the Arts*. Manchester: Manchester University Press.

McKitterick, R. (1999) *The New Cambridge Medieval History*. Cambridge.

Maggi, A. (2003) *John Soane and the Wooden Bridges of Switzerland: Architecture and the Culture of Technology from Palladio to the Grubenmanns*. Switzerland: Accademia di Architettura.

Mainstone, R.J. (1988) *Hagia Sophia: Architecture, Structure and Liturgy of Justinians's Great Church*. London: Thames & Hudson.

Martienssen, H. (1976) *The Shapes of Structure*. Oxford.

Martin, H.C. (1966) *Introduction to Matrix Methods of Structural Analysis*. New York: McGraw-Hill.

Massie, R.K. (1992) *Dreadnought: Britain, Germany and the Coming of the Great War V.1*. London: Jonathan Cape.

Mawson, B. and Lark, R. (1999) Newport Transporter Bridge. *Structural Engineer*, 77.16

Milsom, J. (1975) *Russian Tanks, 1900–1970: The Complete Illustrated History of Soviet Armoured Theory and Design*. New York: Galahad Books.

Monk, R. (1991) *Ludwig Wittgenstein. The Duty of Genius*. London: Vintage.

Monk, R. (1996) *Bertrand Russell*. Free Press.

Morgan, A.G. (1971) *Dams and Other Disasters: A Century of the Army Corps of Engineers in Civil Works*. Boston, MA: Porter Sargent Publisher.

Morrison, J. et al. (1986) *The Athenian Trireme*. Cambridge.

Morrison, J. et al. (2002) *The Age of the Galley*. Conway.

Morrisey, J. (2002) *The Genius in the Design*. Morrow.

Morrissey, J, (2005) *The Genius in the Design: Bernini, Borromini, and the Rivalry That Transformed Rome*. New York: Harper Collins.

Mostafavi, M. (2002) *Structure as Space: Architecture and Engineering in the Work of Jürg Conzett*. Cambridge, MA: MIT Press.

Müller, W. U. and Gunther, V. (1984). *Atlas zur Baukunst*. Stuttgart: Dt. Bücherbund.

Murray, R. (2003) *The Saga of Sydney Opera House: The Dramatic Story of the Design and Construction of the Icon of Modern Australia*. London: Routledge.

# Bibliography

Nabokov, V. (1981) *Lectures on Russian Literature* ed. Fredson Bowers. New York: Harcourt Brace Jovanovich.

Needham, J. (1954) *Science and Civilisation in China*. Cambridge.

Nesbit, M. (2000) *Their Common Sense*. London: Black Dog Publishing.

Newby, F. (2001) *Early Reinforced Concrete – Studies in the History of Civil Engineering*. Aldershot: Ashgate Publishing.

Nicholls, J. & N., and Clute, L. (1993) *The Encyclopedia of Science Fiction*. New York: St. Martin's Press.

Norberg-Schulz, C. (1985) *Late Baroque & Rococo Architecture*. Electa/Rizzoli.

Nordenson, G. ed. (2008) *Seven Structural Engineers: The Felix Candela Lectures*. New York: Museum of Modern Art.

Norwich, J. (1998) *A Short History of Byzantium*. Vintage.

Norwich, J.J. (1989) *A History of Venice*. London: Vintage.

Otto, F. (1995) *Finding Form: Towards an Architecture of the Minimal*. Fellbach, Germany: Edition Axel Menges.

Outram, D. (1984) *Georges Cuvier: Vocation, Science and Authority in Post-revolutionary France*. Manchester: Manchester University Press.

Pacey, A. (1992) *The Maze of Ingenuity*. Cambridge, MA: The MIT Press.

Padovan, R. (1999) *Proportion*. Taylor & Francis.

Pallottino, Massimo et.al. (eds.) (1968) *Encyclopedia of World Art Vols XII Renaissance to Shahn*. New York: McGraw Hill.

Panofsky, E. (1951) *Gothic Architecture and Scholasticism*. Archabbey.

Panofsky. E. ([1955] 1983) *Meaning in the Visual Arts*. Chicago: University of Chicago Press.

Parkinson, R. (1999) *Cracking Codes: The Rosetta Stone and Decipherment*. London: British Museum Press.

Pawley, M. (1999) Konrad Wachsmann: the greatest architect of the twentieth century. *Architects Journal* 2.12.

Pearce, F. (2005) *Deep Jungle*. Transworld.

Pearson, A. (2002) *The Roman Shore Forts*. The History Press.

Pearson, M.P. (1993) *Bronze Age Britain*. London: B.T. Batsford.

Perez-Gomez, A. and Pelletier, L. (1997) *Architectural Representation and the Perspective Hinge*. Cambridge, MA: The MIT Press.

Peters, T.F. (1987) *Transitions in Engineering. Guillaume Henri Dufour and the Early 19th Century Cable Suspension Bridges*. Basel: Birkhauser Verlag.

Petroski, H. (1994) *Design Paradigms: Case Histories of Error and Judgment in Engineering*. Cambridge: Cambridge University Press.

Pevsner, N. (1969) *Ruskin and Viollet-le-Duc: Englishness and Frenchness in the Appreciation of Gothic Architecture*. London: Thames and Hudson.

Pevsner, N. (1991) *An Outline of European Architecture*. London: Penguin.

Picon, A. (1997) *Art De L'Ingenieur*. Paris: Centre Georges Pompidou Service Commercial.

Pippard, A.J.S. and Baker, J.F. (1957) *The Analysis of Engineering Structures*. Arnold.

Plowden, D. (1974) *Bridges, The Spans of North America*. New York: Viking Press.

Pocock, J. (1999) *Barbarism and Religion*. Cambridge.

Polmar,, N. and Moore, K. J. (2003) *Cold War Submarines: The Design and Construction of U.S. and Soviet Submarines, 1945–2001*. Dulles, Virginia: Potomac Books Inc.

Popper, K. ed. (1998) *The World of Parmenides: Essays on the Presocratic Enlightenment*. London: Routledge.

Rabun, J.S. (2000) *Structural Analysis of Historic Buildings: Restoration, Preservation, and Adaptive Reuse Applications for Architects and Engineers*. New York: John Wiley & Sons.

Rapley, J. (2003) *The Britannia and Other Tubular Bridges: And the Men Who Built Them*. Gloucestershire: NPI Media Group.

Rapley, J. (2006) *Thomas Bouch: The Builder of the Tay Bridge*. Gloucestershire: The History Press.

Rastorfer, D. (2000) *Six Bridges: The Legacy of Othmar H. Ammann*. Connecticut: Yale University Press.

Ravetz, J. (1995) *Scientific Knowledge and It's Social Problems*. Transaction.

Reiss, T. (1997) *Knowledge, Discovery and Imagination in Early Modern Europe: The Rise of Aesthetic Rationalism*. Cambridge: Cambridge University Press.

Resnikoff, H.L. and Wells, R.O. (1973) *Mathematics in Civilisation*. New York: Dover.

Rolt, L. T. C. (2000) *Victorian Engineering*. London: Penguin.

Rondelet, J-B. (1817) *Traité théorique et pratique de l'art de bâtir*. Paris: Rondelet.

Rosa, J. (2008) *A Constructed View: The Architectural Photography of Julius Shulman*. New York: Rizzoli.

Rosenbaum, A. et.al. (1994) *Works in Progress*. California: Pomegranate Communications.

Ross, G. M. (1984) *Leibniz (Past Masters Series)*. Oxford: Oxford Paperbacks.

Ryall, M.J. (1996) Management of stresses in concrete structures using an instrumented hand inclusion technique, *Structural Engineer*, 74 (15) 6 August: 255–260.

Rybczynski, W. (2001) *One Good Turn*. Scribner.

Said, E.W. (2003) *Orientalism: Western Conceptions of the Orient*. London: Penguin Classics.

Sato, H. and Nakahara, Y. (2000) *The Complete Japanese Joinery*. Washington: Hartley and Marks Publishers.

Schittich, C. et al. (1999) *Glass Construction Manual*. Basel: Birkhauser Verlag.

Scott, R. (2001) *In the Wake of Tacoma: Suspension Bridges and the Quest for Aerodynamic Stability*. Virginia: American Society of Civil Engineers.

Seike, K. (1977) *The Art of Japanese Joinery*. New York: Weatherhill.

Sharratt, M. (1996) *Galileo: Decisive Innovator*. Cambridge: Cambridge University Press.

Shute, N. (1954) *Slide Rule: Autobiography of an Engineer*. London: Heinemann.

Singer, C. (1990) *A History of Technology: v. 2*. Oxford: Oxford University Press.

Silver, C. (2007) *Renkioi: Brunel's Forgotten Crimean War Hospital*. Sevenoaks: Valonia Press.

Singer, C. et al. (1954) *A History of Technology*. Oxford.

Sjovold, T. (1979) *The Viking Ships in Oslo*. Universitetets Oldsaksmling.

Skulski, J. (1994) *The Heavy Cruiser Takao (Anatomy of the Ship Series)*. Annapolis, Maryland: US Naval Institute Press.

Smith, R.K. (1965) *The Airships Akron and Macon (Flying Aircraft Carriers of the United States Navy)*. Annapolis, Maryland: United States Naval Institute.

Sobel, D. (2005) *Longitude*. London: Harper Perennial.

Sober, E. (1975) *Simplicity*. Oxford.

Sokoloff, V. (1982) *Ships of China*. Patrocinador.

Solomon, S. (2000) *Louis I. Kahn's Trenton Jewish Community Center: Building Studies 6*. Princeton, NJ: Princeton Architectural Press.

Southern, R. (1961) *The Making of the Middle Ages*. Yale.

Speller, E. (2003) *Following Hadrian: A 2nd-Century Journey through the Roman Empire*. London: Headline.

Spindler, K. (2001) *The Man in the Ice*. Phoenix.

Starr, C.G. (1989) *The Influence of Sea Power on Ancient History*. Oxford: Oxford University Press.

Steele, J. (2002) *Eames House: Charles and Ray Eames (Architecture in Detail)*. London: Phaidon Press.

Steer, J. (1980) *A Concise History of Venetian Painting*. Thames & Hudson.

Stern, F. (2001) *Einstein's German World*. Princeton, NJ: Princeton University Press.

Stierlin, H. (1983) *Encyclopaedia of World Architecture*. London: Macmillan.

Stierlin, H. (2002) *The Roman Empire: From the Etruscans to the Decline of the Roman Empire*. Cologne: Taschen.

Stoye, J. (2001) *The Siege of Vienna*. Edinburgh: Birlinn.

Suetonius Tranquillus. (1957) *The Twelve Caesars,* trans. Robert Graves. London: Penguin.

Sutter, J. and Spenser, J. (2006) *747: Creating the World's First Jumbo Jet and Other Adventures from a Life in Aviation*. London: Harper Collins.

Taplin, O. (ed) (2000) *Literature in the Roman World*. Oxford: Oxford University Press.

Terzaghi, K. (1943) *Theoretical Soil Mechanics*. New York: Wiley.

Thom, A. (1971) *Megalithic Lunar Observatories*. Oxford: Oxford University Press.

Thomée, V. (1999) From finite differences to finite elements: a short history of numerical analysis of partial differential equations, *Journal of Computational and Applied Mathematics Archive 1–2*,

# Bibliography

March 2001 (received 1999, published 2001), http://www.math.chalmers.se/Math/Research/Preprints/ 1999/21.ps.gz.

Thompson, D.W. (1992) *On Growth and Form* ed. J. T. Bonner. Cambridge: Cambridge University Press.

Timoshenko, S. (1983) *History of the Strength of Materials*. Dover.

Tinniswood, A. (2002) *His Invention So Fertile: A Life of Christopher Wren*. London: Pimlico.

Torroja, E. (1958) *Philosophy of Structures*. Transl. by J.J. Polivka and Milos Polivka. California: University of California Press.

Totman, C. (2005) *A History of Japan*. Malden: Blackwell.

Trask, W. (1997) *History of My Life by Giacomo Casanova*. Everyman's Library.

Tyldesley, J. (2003) *Pyramids: The Real Story behind Egypt's Most Ancient Monuments*. Penguin.

Underhill, H.A. (1952) *Deep Water Sail*. Glasgow: Brown Son & Ferguson.

UNESCO World Heritage (2005) *29COM 8B.38 – Nominations of Cultural Properties to the World Heritage List (Le Havre, the City Rebuilt by Auguste Perret)*, http://whc.unesco.org/en/list/1181/documents/.

UNESCO World Heritage (2006) *30COM 8B.49 – Nominations of Cultural Properties to the World Hertage List (Vizcaya Bridge)*, http://whc.unesco.org/en/decisions/1011.

Van Dulken, S. (2001) *Inventing the Twentieth Century: 100 Inventions that Shaped the World*. London: British Library.

Vaughan, A. (1991) *Isambard Kingdom Brunel: Engineering Knight Errant*. London: John Murray.

Verne, J. (1970) *Twenty Thousand Leagues Under the Sea*. Norton.

Verner, M. (2002) *The Pyramids: Their Archaeology and History*. Atlantic.

Washington Government Printing Office (1947) *Building the Navy's Bases in World War II: History of Yards and Docks and the Civil Engineering Corps 1940–1946 Vol. 1*. WGPO.

Watson, A. (2003) *Aurelian and the Third Century*. London: Routledge.

Whyte, L.L. (1951) *Aspects of Form: a Symposium on Form in Nature and Art*. Aldershot: Lund Humphries.

Wigelsworth, J. (2006) *Science and Technology in Medieval European Life*. Greenwood.

Willetts, R. (2004) *The Civilisation of Ancient Crete*. Phoenix.

Wittkower, R. (1973) *Architectural Principles in the Age of Humanism*. London: Academy Editions.

Wykeham, P. (1980) *Santos-Dumont*. New York: Arno.

Yang, J-N. and Giannopoulos, N. (1979) Active Control and Stability of Cable-Stayed Bridges. *ASCE. Journal of the Engineering Mechanics Division* 105 (4) July/August 1979, 677–694.

Yates, F.A. (1972) *The Rosicrucian Enlightenment*. London: Routledge & Kegan Paul.

Zamoyski, A. (2005) 1812: *Napoleon's Fatal March on Moscow*. Harper.

# Index

Page numbers in *italics* denotes an illustration.

# Index

Printed and bound by CPI Group (UK) Ltd, Croydon, CR0 4YY

01/11/2024

01782611-0001